Contents at a Glance

Table of Contents

Introduction

*I*t may be a smartphone, but it makes you feel dumb. Don't worry: You aren't alone. As technology leaps ahead, it often leaves mortal humans behind. You paid good money for your phone — why not use all of its features?

This book makes the complex subject of Android phones understandable. It's done with avuncular care and gentle handholding. The information is friendly and informative, without being intimidating. And yes, ample humor is sprinkled throughout the text to keep the mood light.

About This Book

I implore you: Do not read this book from cover to cover. This book is a reference. It's designed to be used as you need it. Look up a topic in the table of contents or the index. Find something about your phone that vexes you or something you're curious about. After getting the answer, get on with your life.

Every chapter in this book is written as its own self-contained unit, covering a specific Android phone topic. The chapters are further divided into sections representing tasks you perform with the phone or explaining how to get something done. Sample sections in this book include

- ✔ Typing without lifting a finger
- ✔ Making a conference call
- ✔ Dealing with a missed call
- ✔ Uploading a picture to Facebook
- ✔ Recording video
- ✔ Creating a mobile hotspot
- ✔ Flying with your phone
- ✔ Extending battery life

Every section explains a topic as though it's the first one you read in this book. Nothing is assumed, and everything is cross-referenced. Technical terms and topics, when they come up, are safely shoved to the side, where they're easily avoided. The idea here isn't to learn anything. This book's philosophy is to help you look it up, figure it out, and move on.

How to Use This Book

This book follows a few conventions for using your phone, so pay attention!

The main way to interact with the typical Android phone is by using its *touchscreen,* which is the glassy part of the phone as it's facing you. The physical buttons on the phone are called *keys.* These items are discussed and explained in Part I of this book.

Various ways are available to touch the screen, which are described in Chapter 3.

Chapter 4 covers typing text on an Android phone, which involves using something called the *onscreen keyboard.* When you tire of typing, you can dictate your text. It's all explained in Chapter 4.

This book directs you to do things on your phone by following numbered steps. Every step involves a specific activity, such as touching something on the screen; for example:

1. Choose Downloads.

This step directs you to tap the text or item on the screen labeled Downloads. You might also be told to do this:

1. Tap Downloads.

Because this book covers a variety of phones, alternative commands may be listed. One of them is bound to match something on your phone, or at least be close to what you see:

1. Tap the My Downloads action or the Downloads action.

 Various phone settings are turned off or on, as indicated by a box next to the setting. Touch this box on the screen to add or remove a check mark. When the check mark appears, as shown in the margin, the option is on; otherwise, it's off.

Some settings feature a master control, which looks like the on/off switch shown in the margin. Slide the button to the right to activate the switch, turning on a phone feature. Slide the button to the left to disable the feature.

Foolish Assumptions

Even though this book is written with the gentle handholding required by anyone who is just starting out or is easily intimidated, I have made a few assumptions.

I'm assuming that you're still reading the introduction. That's great. It's much better than getting a snack right now or checking to ensure that the cat isn't chewing through the TV cable again.

My biggest assumption: You have an Android phone. It can be any Android phone from any manufacturer supported by any popular cellular service provider in the United States. Because Android is an operating system, the methods of doing things on one Android phone are similar, if not identical, to doing things on another Android phone. Therefore, one book can pretty much cover the gamut of Android phones.

Android has versions. This book was updated to cover the current Android release, 5.1, known as Lollipop. Also addressed are KitKat and Jelly Bean, version numbers 4.3 and 4.4, respectively. These are all similar versions of the operating system, so if your phone has an older version, you should be just fine.

To confirm which Android version your phone has, follow these steps:

1. **At the Home screen, touch the Apps icon.**

2. **Open the Settings app.**

3. **Choose the About Phone item.**

    On some Samsung phones, you need to first tap the General tab atop the screen and then swipe down the screen to find an About Device item. Samsung phones can be a little different from other Android phones, and those differences are highlighted throughout this tome.

4. **Look at the item titled Android Version.**

    The version number is listed, such as 5.1.

Don't fret if these steps confuse you: Review Part I of this book, and then come back here. (I'll wait.)

More assumptions: You don't need to own a computer to use your Android phone. If you have a computer, great. The Android phone works well with both PCs and Macs. When phone and computer cross paths, you'll find directions for both PC and Mac.

Finally, this book assumes that you have a Google account. If you don't, find out how to configure one in Chapter 2. Having a Google account opens up a slew of useful features, information, and programs that make using your phone more productive.

How This Book Is Organized

This book has been sliced into five parts, each of which describes a certain aspect of the typical Android phone or how it's used.

Part I: Getting Started with Your Android Phone

Part I serves as an introduction to your Android phone. Chapters cover setup and orientation and familiarize you with how the phone works. This part is a good place to start — plus, you discover things in this part that aren't obvious from just guessing how the phone works.

Part II: Reach Out and Touch Someone

This part of the book is about communication. It starts with the obvious functions of making a phone call, dealing with voicemail, and using the phone's address book. Also covered are communication tools, such as text messaging, email, using the web, and social networking opportunities.

Part III: Amazing Phone Feats

Part III explores the non-phone things your phone can do. For example, your phone can find locations on a map, give you verbal driving directions, take pictures, shoot videos, play music, play games, and do all sorts of wonderful things that no one would ever think a phone can do. The chapters in this part of the book get you up to speed on those activities.

Part IV: Nuts and Bolts

The chapters in Part IV discuss a slate of interesting topics, from connecting the phone to a computer, using Wi-Fi and Bluetooth networking, and taking the phone overseas and making international calls to adding security, working with apps, and personalizing your phone. Also addressed are the necessary chores of maintenance and troubleshooting.

Part V: The Part of Tens

Finally, this book ends with the traditional *For Dummies* Part of Tens, where every chapter lists ten items or topics. For your Android phone, the chapters include tips, tricks, shortcuts, and things to remember.

Icons Used in This Book

This icon flags useful, helpful tips or shortcuts.

This icon marks a friendly reminder to do something.

This icon marks a friendly reminder *not* to do something.

This icon alerts you to overly nerdy information and technical discussions of the topic at hand. Reading the information is optional, though it may win you a pie slice in *Trivial Pursuit*.

Where to Go from Here

Thank you for reading the introduction. Few people do, and it would save a lot of time and bother if they did. Consider yourself fortunate, although you probably knew that.

Your task now: Start reading the rest of the book — but not the whole thing, and especially not in order. Observe the table of contents and find something that interests you. Or look up your puzzle in the index. When these suggestions don't cut it, just start reading Chapter 1.

My email address is dgookin@wambooli.com. Yes, that's my real address. I reply to every email I receive, especially when you keep your question short and specific to this book. Although I enjoy saying "Hi," I cannot answer technical support questions, resolve billing issues, or help you troubleshoot your phone. Thanks for understanding.

My website is `www.wambooli.com`. This book has its own page on that site, which you can check for updates, new information, and all sorts of fun stuff. Visit often:

`www.wambooli.com/help/android/phones`

Bonus information for this title can be found online. You can visit the publisher's website to find an online Cheat Sheet at

`www.dummies.com/cheatsheet/androidphones`

Supplemental online material has been created for this book. That supplemental stuff can be found at

`www.dummies.com/extras/androidphones`

Enjoy this book and your Android phone!

Part I
Getting Started with Your Android Phone

In this part . . .

- ✔ Get started with Android phones.
- ✔ Work through the setup of your Android phone.
- ✔ Learn how to use your Android phone.
- ✔ Discover parts of the Android phone.

Hello, Phone!

In This Chapter

▶ Freeing the phone from its box

▶ Setting up the phone

▶ Charging the battery

▶ Recognizing phone features

▶ Obtaining accessories

▶ Taking the phone with you

▶ Keeping the phone in one place

*I*t may have a funky name, like a character in a science fiction novel or a sports hero. Or it can simply be a fancy number, perhaps with the letter *X* thrown in to make it cool. No matter what, the phone you own is an *Android* phone because it runs the Android operating system. The adventure you're about to undertake begins with removing the thing from the box and getting to know your new smartphone.

Liberation and Setup

The phone works fastest when you remove it from its box. How you liberate it is up to you. I prefer to gingerly open the box, delicately lifting the various flaps and tenderly setting everything aside. I even savor the industrial solvent smell. If you prefer, you can just dump everything on the tabletop. But be careful: Your phone may be compact, but it's not indestructible.

Several useful items are found loitering inside your Android phone's box. Some of them are immediately handy, and others you should consider saving for later. Even if you've already opened the box and spread its contents across the table like some sort of tiny yard sale, take a few moments to locate and identify these specific items:

- The phone itself, which may be fully assembled or in pieces
- Papers, instructions, a warranty, and perhaps a tiny, useless *Getting Started* pamphlet
- The phone's battery, which might already be installed inside the phone
- The phone's back cover, which might already be on the phone
- The charger/data cable or USB cable
- The charger head, which is a wall adapter for the charger/data cable
- Other stuff, including the SIM card, SIM card removal tool, headphones, a carrying case, or other goodies

If anything is missing or appears to be damaged, immediately contact the folks who sold you the phone.

I recommend keeping the instructions and other information as long as you own the phone: The phone's box makes an excellent storage place for that stuff — as well as for anything else you don't plan to use right away.

See the later section "Adding accessories" for a description of various goodies available for the typical Android phone.

Android phone buying tips

When buying a phone, first look at a cellular provider, and then determine which phones are available and suit your purpose.

Finding a cellular provider is all about coverage: Can you get a signal everywhere you need one? Despite the boasts, not every cellular provider offers full data coverage. The true test is to ask people who frequent your same locations which services they use and whether they're happy with the coverage.

All Android phones offer similar features and a vast array of apps. Start looking for a phone by finding something that feels good in your hands. Some people like smaller, compact phones that fit easily in a pocket or purse. Others prefer the large-format *(phablet)* phones, which offer larger screens.

Check the phone's display, not by reviewing the fancy technical jargon but by looking at it with your own eyes. View some photos on the phone to see how good they look.

If you need more storage, look for a phone that offers a removable MicroSD card. This feature is becoming rather rare, so if you want it, you may have to look for it.

Camera resolution isn't vital, but if your Android phone is your only digital camera, getting a high-resolution rear camera is a plus.

Beyond these basic items, most Android phones are drearily similar. To work best with this book, ensure that your phone uses the Android operating system and can access and use the Google Play Store. Some low-price, bargain phones restrict your purchases to the manufacturer's own app store. That's not a good thing.

Phone Assembly

Most Android phones come fully assembled. If not, the folks at the store most likely put everything together for you. When you're on your own, some setup may be required. This process involves installing the SIM card and inserting the battery. Directions that come with the phone assist you.

Don't worry about the assembly process being overly complex; if you're good with Legos, you can put together an Android phone.

- ✔ The phone ships with a clingy plastic cover over its screen, back, or sides. The plastic thingies might tell you where to find various features or how to install the battery. Please remove all clingy plastic sheets. Also check the phone's rear camera to confirm that you've removed the plastic thingy from its lens.

- ✔ The current trend is for Android phones not to feature a removable battery. If your phone has a removable battery, refer to the included instructions for details on inserting and removing the battery.

Installing the SIM card

A *SIM card* identifies your phone on a digital cellular network. Before you can use the phone, the SIM card must be installed. The only time you as a mere mortal need to do this is when you purchase the phone independently of a cellular provider. Otherwise, the kind people at the Phone Store install the SIM card.

When you do need to install the SIM card, follow these steps when the phone is turned off:

1. **Pop the SIM card out of the credit-card-size holder.**

2. **Insert the SIM card into the SIM card slot.**

   On some phones, the SIM card is inserted internally. In that case, remove the phone's back cover and, if necessary, remove the battery to access the SIM card slot.

   On other phones, the SIM slot is found on the device's outer edge. You must open the tiny SIM slot cover and insert the SIM card into the slot. This procedure may require a special pointy-thing tool that came with your phone; insert the pointy-thing into the hole by the SIM slot cover to open the cover. Insert the SIM card directly into the slot, or first place the card into a SIM card carrier and then insert it.

   The SIM card is shaped in such a way that it's impossible to insert improperly. If the card doesn't slide into the slot, reorient the card and try again.

3. **If necessary, replace the battery and the phone's back cover.**

   You're done.

The good news is that you seldom, if ever, need to remove or replace a SIM card.

SIM stands for subscriber identity module. SIM cards are required for GSM cellular networks as well as for 4G LTE networks.

Installing a MicroSD card

A few Android phones offer removable storage in the form of a MicroSD card. If your phone offers this feature, go out and obtain a MicroSD card to take advantage of the extra storage.

To insert the MicroSD card, heed these directions:

1. **Locate the slot into which you stick the MicroSD card.**

   The slot is labeled as shown in Figure 1-1. It is not the same as the SIM card slot.

Lift here

Figure 1-1: Opening the MicroSD card hatch.

2. **Flip open the teensy hatch on the MicroSD card slot.**

   Insert your thumbnail into the tiny slot on the hatch. Flip the hatch outward. It's attached on one end, so it may not completely pop off.

3. **Insert the MicroSD card into the slot.**

   The card goes in only one way. If you're fortunate, a little outline of the card illustrates the proper orientation. If you're even more fortunate, your eyes will be good enough to see the tiny outline.

You may hear a faint clicking sound when the card is fully inserted. If not, use the end of a paperclip or your fingernail to fully insert the card.

✔ It's okay to insert the MicroSD card while the phone is on.

✔ If the phone is on, a prompt appears on the touchscreen detailing information about the card. The card is available instantly for use. If not, the card may need to be formatted. See Chapter 18 for details.

✔ Some older Android phones may feature internal MicroSD card slots. In that case, you must remove the phone's back cover to access and install the card. You might even have to remove the battery to get at the card.

✔ I've never seen an Android phone come with a MicroSD card. If your phone can use such a card, obtain one at any computer or office supply store. They're cheaper if you order them on the Internet.

✔ A MicroSD card comes in a capacity rated in gigabytes (GB), just like most media storage or memory cards. Common MicroSD card capacities are 8GB, 16GB, 32GB, and higher. The maximum size allowed in your phone depends on its design. The side of the phone's box lists compatible capacities.

Removing the MicroSD card

To remove the MicroSD card, follow these steps:

1. **Turn off the phone.**

   It's possible to remove the card while the phone is on, and directions are offered in Chapter 18. For now, ensure that the phone is off. Specific power-off directions are found in Chapter 2.

2. **Open the little hatch covering the MicroSD card slot.**

   Refer to the preceding section.

3. **Using your fingernail or a bent paperclip, gently press the MicroSD card inward a tad.**

   The MicroSD card is spring-loaded, so pressing it in pops it outward.

4. **Pinch the MicroSD card between your fingers and remove it completely.**

After you've removed the card, you can continue using the phone. It works just fine without a MicroSD card.

✔ A MicroSD card is teensy! If you remove it from your phone, keep it in a safe place where you won't lose it. Never stick the MicroSD card into your ear.

✔ You can purchase MicroSD card adapters to allow the card's data to be read by a computer, by either a standard SD memory slot or the USB port.

✔ Refer to Chapter 18 for more information on phone storage.

Charge the Battery

The phone's battery may have enough oomph in it to run the setup-and-configuration process at the Phone Store. If so, count yourself lucky. Otherwise, you need to charge the phone's battery. Don't worry about flying a kite and waiting for a lightning storm. Instead, follow these steps:

1. **If necessary, assemble the charging cord.**

   Connect the charger head (the plug thing) to the USB cable that comes with the phone. They connect in only one way.

2. **Plug the charger head and cable into a wall socket.**

3. **Plug the phone into the USB cable.**

   The charger cord plugs into the micro-USB connector, found at the phone's bottom. The connector plugs in only one way.

As the phone charges, you may see a charging battery graphic on the touch-screen, or a notification lamp on the phone's front side may glow. Such activity is normal.

The phone may turn on when you plug it in for a charge. That's okay, but read Chapter 2 to find out what to do the first time the phone turns on. You also may need to contact your cellular provider for additional setup instructions the first time you turn on the phone.

- ✔ I recommend fully charging the phone before you use it.

- ✔ You can use the phone while it's charging, although the phone won't turn on when the battery is too low.

- ✔ The phone also charges itself whenever it's plugged into a computer by way of a USB cable. The computer must be on for charging to work. Some phones may charge only when plugged into a powered USB port, such as those found directly on the computer console.

- ✔ Cell phones charge more quickly when plugged into the wall than into a computer's USB port or a car adapter.

- ✔ Unlike the old NiCad batteries, you don't need to worry about fully discharging your phone before recharging it. If the phone needs a charge, even when the battery is just a little low, feel free to do so.

- ✔ Some Android phones can be charged wirelessly. See the later section "Adding accessories."

- ✔ Also see Chapter 23 for battery and power management information.

Android Phone Orientation

No one told the first person to ride a horse which way to sit. Some things just come naturally. Your Android phone most likely isn't one of those things. It requires a special introduction and orientation.

Finding things on your phone

I think it's cute when people refer to things that they can't name as a *doodad* or *thingamabob*. Cute, but inaccurate. Take a gander at Figure 1-2, which illustrates common items found on the front and back of a typical Android phone.

Not every item shown in the figure may be in the exact same spot on your phone. For example, the Power/Lock key might be found on the top of the phone, not the side.

The terms referenced in Figure 1-2 are the same as the terms used elsewhere in this book and in whatever scant Android phone documentation exists. Here are the highlights:

Power/Lock key: The Power/Lock key does more than just turn on or off the phone, which is why it's the Power/Lock key and not the On/Off button.

Volume key: The phone's volume control is two buttons in one. Press one end of the key to set the volume higher; the other end sets the volume lower. This key might also be used to control the phone's camera, as covered in Chapter 13.

Touchscreen display: The main part of the phone is its *touchscreen* display. It's a see-touch thing: You look at the display and also touch it with your fingers to control the phone. That's where it gets the name *touch*screen.

Front camera: The phone's front-facing camera is found above the touchscreen. It's used for taking self-portraits as well as for video chat.

Speaker(s): The primary phone speaker is located top center on the phone. One or more additional speakers might also be found on the phone's bottom edge or backside.

Microphone: Somewhere below the touchscreen, you'll find the microphone. It's tiny, about the diameter of a pin. Don't stick anything into the hole! A second, noise-canceling microphone might also be found on the back of the phone.

Volume key   Power/Lock key

Front camera   Speaker   Rear camera   LED flash   Headphone jack

ANDROID PHONE

Touchscreen display   Microphone   USB/Power connector   Rear speaker

Front   Back

Figure 1-2: Your phone's face and rump.

Headphone jack: Somewhere on the phone's edge you'll find a hole where you can connect standard headphones. Its location may not always be on top, as shown earlier, in Figure 1-2.

Rear camera: The rear camera is found on the phone's back. It may be accompanied by one or two LED flash gizmos.

USB/Power connector: Use this important connector to attach the phone's USB cable. That cable is used both to charge the phone and communicate with a computer. This connector might be on the side of the phone, but more commonly it's found bottom center.

Take a moment to locate all the items mentioned in this section, as well as shown in Figure 1-2, on your own phone. It's important that you know where they are.

✔ Additional items you might want to locate on your phone, items not illustrated in Figure 1-2, include the SIM card cover and media card slot. Use of these features is covered earlier in this chapter.

✔ Some phones, such as those in the Samsung Galaxy line, feature a physical button called the Home key. This key is found below the touch-screen.

✔ The Galaxy Note line of phones features a pointing device, in the form of a digital stylus called an *S Pen*. It docks at the phone's bottom edge.

✔ The phone's microphone picks up your voice, loud and clear. You don't need to hold the phone at an angle for the microphone to work.

Using earphones

You don't need to use earphones to get the most from your Android phone, but it helps! If the nice folks who sold you the phone tossed in a pair earphones, that's wonderful! If they didn't, well then, they weren't so nice, were they?

The most common type of cell phone earphones are the *earbud* style: You set the buds into your ears. The sharp, pointy end of the earphones, which you don't want to stick into your ear, plugs into the phone's headphone jack.

Between the earbuds and the sharp, pointy thing, you might find a doodle on which a button sits. The button can be used to mute the phone or to start or stop the playback of music when the phone is in its music-playing mode.

You can also use the doodle button to answer the phone when it rings.

A teensy hole on the doodle serves as a microphone. The mic allows you to wear the earphones and talk on the phone while keeping your hands free. If you gesture while you speak, you'll find this feature invaluable.

✔ Some phones come with a set of earphones. If yours did not, you can purchase a set at any electronics store where the employees wear name tags. Ensure that the headset features a microphone.

- ✔ Some headsets feature extra doodle buttons, such as Forward and Reverse buttons. Though the headset may work fine with your phone, the extra buttons probably don't do anything specific.

- ✔ Be sure to fully insert the earphone connector into the phone. The person you're talking with can't hear you well when the earphones are plugged in only part of the way.

- ✔ You can also use a Bluetooth headset with your phone, to listen to a call or some music. See Chapter 17 for more information on Bluetooth.

- ✔ Fold the earphones when you put them away, as opposed to wrapping them in a loop. Put the earbuds and connector in one hand, and then pull the wire straight out with the other hand. Fold the wire in half and then in half again. You can then put the earphones in your pocket or on a tabletop. By folding the wires, you avoid creating something that looks like a wire ball of Christmas tree lights.

Adding accessories

Beyond earphones, you can find an entire Phone Store full of accessories and baubles that you can obtain for your Android phone. The variety is seemingly endless, and the prices, well, they ain't cheap. Here are some of your choices:

Phone case: Protect your phone by getting it a jacket, one that further expresses your individuality.

Screen protectors: These clear plastic sheets adhere to the touchscreen, protecting it from scratches and other abuse. Ensure that you get screen protectors designed specifically for your phone.

Belt clip: To sate your envy of Batman's utility belt, consider getting a fine leatherette or Naugahyde phone case that you can quickly attach to your belt.

Car charger: This gizmo connects your phone to the car's 12-volt battery source. If you're over 40, the power source was once known as a cigarette lighter.

Car mount: This device holds your Android phone so that you can easily see it while driving. It makes for easier access, although these things are forbidden in some states. See Chapter 17 for information on using Bluetooth in your car for hands-free phone operations.

MicroSD Card: When your phone supports this type of removable storage, consider buying this memory card. See the earlier section "Installing a MicroSD card."

Dock: This is a heavy base into which you can set your phone. Some docks are simple cradles that prop up the phone for easy bedside viewing. Others are more sophisticated, offering USB connectors and maybe even a real keyboard.

Wireless charger: Not every phone can be charged wirelessly, but if yours can, definitely get a wireless charger. Simply set your phone on the pad or prop it up in the dock. The phone's battery starts magically recharging.

Screencasting dongle: This accessory connects to an HDTV or computer monitor. Once configured, it allows you to cast the phone's screen onto the larger-screen device. It's ideal for watching movies, Netflix, or YouTube videos or for enjoying music. Google's Chromecast is an example of a screencasting gizmo.

A Home for Your Phone

Homes built in the 1930s and 1940s often featured a special hole in the wall called a *phone cubby*. It was the shrine in which you would find the home's hard-wired phone. You should treat your Android phone with similar reverence and respect.

Toting your Android phone

The compactness of the modern cell phone makes it perfect for a pocket or even the teensiest of party purses. And its well-thought-out design means you can carry your phone in your pocket or handbag without fearing that something will accidentally turn it on, dial Mongolia, and run up a heck of a cell phone bill.

Your Android phone most likely features a proximity sensor, so you can even keep the phone in your pocket while you're on a call. The proximity sensor disables the touchscreen, which ensures that nothing accidentally gets touched when you don't want it to be touched.

- ✔ Though it's okay to place the phone somewhere when you're making a call, be careful not to touch the phone's Power/Lock key (refer to Figure 1-2). Doing so may temporarily enable the touchscreen, which can hang up a call, mute the phone, or do any of a number of undesirable things.

- ✔ Don't forget that you've placed the phone in your pocket, especially in your coat or jacket. You might accidentally sit on the phone, or it can fly out when you peel off your coat. The worst fate for any cell phone is to take a trip through the wash. I'm sure your phone has nightmares about it.

Storing the phone

I recommend that you find a single place for your phone when you're not taking it with you: on top of your desk or workstation, in the kitchen, on the nightstand — you get the idea. Phones are as prone to being misplaced as are your car keys and glasses. Consistency is the key to finding your phone.

Then again, your phone rings, so you can always have someone else call your cell phone to help you locate it.

- If available, a docking station makes a handsome, permanent location for your Android phone.

- While working, I keep my phone next to my computer. Conveniently, I have the charger plugged into the computer so that the phone remains plugged in, connected, and charging when I'm not using it.

- Phones on coffee tables get buried under magazines and are often squished when rude people put their feet on the furniture.

- Avoid putting your phone in direct sunlight; heat is bad news for any electronic gizmo.

The On/Off Chapter

*I*t would be delightful if your Android phone were smart enough to pop out of the box, say "Hello," and immediately know everything about you. It doesn't, of course. That introduction is still necessary and it requires some careful attention. It's all part of the initial setup-and-configuration process that happens when you first turn on the phone. And if you haven't yet turned on your phone, this chapter helps with that ordeal as well.

Charging (86%)

Hello, Phone

Modern, technical gizmos lack an on–off switch. Instead, they feature a power button. In the case of your Android phone, the button is called the *Power/Lock* key. This key is used in several ways, some of which may not be obvious or apparent. That's why I wrote this section.

✔ The setup process works more smoothly when you already have a Google, or Gmail, account. If you lack such an account, you're prompted to create an account in the setup process, as described in the next section.

✔ The phone won't start unless the battery is charged. See Chapter 1.

Turning on your phone for the first time

The very first time you turn on an Android phone is a special occasion. That's when you're required to work through the setup-and-configuration process. Don't worry: It needs to be done only once. After that, turn on your phone according to the directions in the next section, "Turning on the phone."

The specifics for the setup and configuration differ depending on the phone's manufacturer as well as on the cellular provider. Odds are pretty good that the people at the Phone Store helped you through the initial setup process. If not, read the generic Android phone setup process outlined in this section, and see the notes at the end of the section for details that may apply to your specific phone.

1. **Turn on the phone by pressing the Power/Lock key.**

   You may have to press it longer than you think. When you see the phone's logo on the screen, you can release the key.

   It's okay to turn on the phone while it's plugged in and charging.

2. **Answer the questions presented.**

   You're asked to select options for some, if not all, of the following items:

   - Select your language
   - Activate the phone on the mobile data network
   - Choose a Wi-Fi network (can be done later)
   - Set the time zone
   - Sign in to your Google account
   - Add other online accounts
   - Set location information

   When in doubt, just accept the standard options as presented to you during the setup process.

   To fill in text fields, use the onscreen keyboard. See Chapter 4 for keyboard information.

   Other sections in this chapter, as well as throughout this book, offer information and advice on these settings. You can't screw up anything at this point; any selection you make can be changed later.

3. **After each choice, tap the Next button or icon.**

   The button might be labeled with the text *Next,* or it may appear as an icon, as shown in the margin.

4. **Tap the Finish button.**

   The Finish button appears on the last screen of the setup procedure.

The good news is that you're done. From this point on, starting the phone works as described in the next section.

After the initial setup, you're taken to the Home screen. Chapter 3 offers more Home screen information, which you should probably read right away, before the temptation to play with your new phone becomes unbearable.

- ✓ You may find yourself asked various questions or prompted to try various tricks when you first start to use your phone. Some of those prompts are helpful, but it's okay to skip some. To do so, tap the OK I GOT IT button or select the Do Not Show Again check box.

- ✓ Additional information on connecting your phone to a Wi-Fi network is found in Chapter 17.

- ✓ Location items relate to how the phone knows its position on Planet Earth. I recommend activating all these items to get the most from your Android phone.

- ✓ It's not necessary to use any specific software provided by the phone's manufacturer or your cellular provider. For example, if you don't want a Samsung account, you don't need to sign up for one; skip that step.

- ✓ By setting up your Google account, you coordinate with your new Android phone whatever information you have on the Internet. This information includes your email messages and contacts on Gmail, appointments on Google Calendar, and information and data from other Google Internet applications.

- ✓ See the later sidebar "Who is this Android person?" for more information about the Android operating system.

Turning on the phone

To turn on your Android phone, press and hold the Power/Lock key. After a few seconds, you feel the phone vibrate slightly and then see the phone's startup animation, logo, or hypnotic brainwashing image. Release the Power/Lock key; the phone is starting.

Eventually, you see the phone's unlock screen. See the later section "Working the screen lock" for what to do next.

When you've encrypted your phone's data, you have to work a screen lock before the device fully starts. See Chapter 21 for details on encrypting your phone's data.

Unlocking the phone

Most of the time, you don't turn your phone off and on. Instead, you lock and unlock it. To unlock and use the phone, press the Power/Lock key. A quick press is all that's needed. The phone's touchscreen comes to life, and you see one of several types of screen lock. Working these locks is covered in the next section.

After you work the screen lock, you next see the Home screen. Chapter 3 covers how to interact with the Home screen.

✔ You can press the Home button to unlock your Samsung phone. The Home button is centered below the touchscreen.

✔ Unlock a Samsung Galaxy Note phone by removing the S Pen.

✔ The phone doesn't unlock for an incoming call; you're able to answer the phone or decline the call. See Chapter 5 for more information on answering, declining, and ignoring incoming calls.

Working the screen lock

The standard Android screen lock is the swipe lock, illustrated in Figure 2-1. Locate a Lock icon on the screen, and then swipe the icon in the direction indicated. Onscreen animation may assist you with working the Swipe lock.

The Swipe lock isn't a difficult lock to pick. If you've added more security, you might see any one of several different screen locks on your phone. Here are the common Android screen locks:

Swipe: The standard screen lock. Swipe your finger on the screen to unlock the device, as illustrated in Figure 2-1. If a screen lock with higher security is set, work it next.

Pattern: Trace a preset pattern over dots on the screen.

PIN: Use the onscreen keyboard to type a number to unlock the device.

Password: Type a password, which can include letters, numbers, and symbols.

None: When this non-lock is selected, the device lacks a screen lock and you can use the phone immediately after pressing the Power/Lock key.

Some phones provide additional, fancy screen locks. Among the variety are the following:

Face Unlock: Look at the phone. In a few moments, your visage is recognized and the phone unlocks.

Figure 2-1: Unlocking Android phones featuring the Swipe lock.

Fingerprint: Brush your finger or thumb over the sensor, such as the Home button on a Samsung Galaxy S6, to unlock the device.

Signature: The Samsung Galaxy Note line of phones lets you use the S Pen to sign your name on the screen. If the signature matches, the phone unlocks.

To use some phone features, you must choose either the PIN or password screen lock, which are considered the most secure. Other locks offer moderate security, and may require a PIN or password as a backup. The swipe and None locks are the least secure.

See Chapter 21 for information on setting the screen lock.

Unlocking and running an app

The phone's lock screen may feature app icons. You'll find them at the bottom of the screen (refer to Figure 2-1), if they're available. Use those icons to both unlock the phone and run the given app.

Who is this Android person?

Just like a computer, your phone has an operating system. It's the main program in charge of all the software (apps) inside the phone. Unlike on a computer, however, Android is a mobile device operating system, designed primarily for use in cell phones and tablets.

Android is based on the Linux operating system, which is also a computer operating system, though it's much more stable and bug-free than Windows, so it's not as popular. Google owns, maintains, and develops Android, which is why your online Google information is synced with the phone.

The Android mascot, shown here, often appears on Android apps or hardware. He has no official name, though most folks call him Andy.

For example, to place a quick phone call, swipe the phone icon up the screen. The phone unlocks and the Phone app appears. Similarly, you can swipe the Camera app icon to snap a quick photo.

✔ Some phones let you customize the lock screen apps, such as the Galaxy Note shown earlier, in Figure 2-1.

✔ When a secure screen lock is set, the phone isn't actually unlocked when the app runs. To do anything else with the phone, you must eventually work the screen lock.

Add More Accounts

Your Android phone can be home to your various online incarnations. That includes your email accounts, online services, social networking, and subscriptions, plus other digital personas. I recommend adding those accounts to your phone as you continue the setup-and-configuration process.

With your phone on and unlocked, follow these steps:

1. Tap the Apps icon.

The Apps icon is found at the bottom of the Home screen. It looks similar to the icon shown in the margin, although it has many variations. See Chapter 3 for the variety.

After tapping the Apps icon, you see the Apps drawer, which lists all apps available on your phone.

2. **Open the Settings app.**

   You may have to swipe the Apps drawer screen a few times, paging through the various icons, to find the Settings app.

   After you tap the Settings icon, the Settings app runs. Then you can configure and set phone options and features.

3. **Choose the Accounts category.**

   On some Samsung phones, first tap the General tab atop the Settings app screen to locate the Accounts category.

   The category may be titled Accounts and Sync on some phones.

   Upon success, you see all existing accounts on your phone, similar to what's shown in Figure 2-2.

Figure 2-2: Accounts listed in the Settings app.

Transferring information from your old phone

Here's one task you don't need to worry about: All the Google information associated with your old phone — or any other Android device, including a tablet — is instantly transferred to your new phone. This information includes contacts, Gmail, events, and other Googly account data. You can even install apps you've previously obtained (free or purchased).

As you add accounts to your phone, the information associated with those accounts is migrated to the device. You might also see media, such as photos, videos, and music transferred, if you use online sharing services to host the media. For information that's not transferred, see Chapter 18, which covers methods for moving files between your phone and other devices.

4. **Tap Add Account.**

The Add Account item is illustrated in Figure 2-2, although it may appear as a plus sign, shown in the margin.

5. **Choose an account from the list.**

For example, to add a Facebook account, choose the Facebook app from the list.

Don't worry if you don't see the exact type of account you want to add. You may have to install a specific app before an account appears. Chapter 16 covers installing new apps on your phone.

6. **Follow the directions on the screen to sign in to your account.**

The steps that follow depend on the account. Generally speaking, you sign in using an existing username and password.

You can continue adding accounts by repeating these steps. When you're done, tap the Home navigation icon to return to the Home screen.

✔ See Chapter 9 for specifics on adding email accounts to your Android phone.

✔ Chapter 11 covers social networking on your phone and offers advice on adding those types of accounts.

Goodbye, Phone

You can dismiss your Android phone in several ways, only two of which involve using a steamroller or raging elephant. The other methods are covered in this section.

Locking the phone

To lock your Android phone, press and release the Power/Lock key. The touchscreen display turns off and the phone is locked.

✔ Your phone will spend most of its time locked. The phone still works while locked; it receives email and plays music and signals alerts and alarms. Phone calls arrive. Yet, while the phone is locked, it doesn't use as much power as it does when the display on.

✔ You can lock the phone while you're making a call. Simply press and release the Power/Lock key. The call stays connected, but the display is disabled.

✔ Locking doesn't turn off the phone.

✔ Refer to Chapter 20 for information on setting the automatic lock time-out value.

Turning off the phone

To turn off your phone, obey these steps:

1. **Press and hold down the Power/Lock key.**

   When you see the Phone Options card, similar to what's shown in Figure 2-3, you can release the key.

2. **Tap the Power Off item.**

   If a confirmation message appears, tap the OK button. The phone turns itself off.

The phone doesn't receive calls when it's turned off. Likewise, any alarms or reminders you've set won't sound when the phone is off.

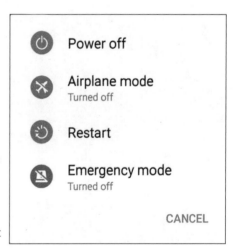

Figure 2-3: The Phone Options card.

✔ Calls received while the phone is off are routed instead to voice mail.

✔ The phone can be charged while it's off.

✔ The Phone Options card might sport more or fewer options than what's shown in Figure 2-3. The bare minimum is the Power Off item. I've also seen Sleep and Kid Mode items, as well as volume and vibration settings.

The Android Tour

ecades ago, the most powerful computers had the most buttons, knobs, and dials. Those room-size computers that ran the 1960 census had hundreds of switches and controls. Back then it would have been easy to assume that a powerful computer of the future would have thousands of switches and controls. The reality is far different.

Today, smart devices like your Android phone are basically bereft of buttons. Despite its seemingly endless potential, your phone is a simple device. The touchscreen is the main control. That's probably not what folks expected back in the 1960s, and it's probably not what you expect today. Therefore, I present this chapter to help you get up to speed on how to operate that advanced gizmo, the Android cell phone.

Andrea Bocelli

Busta Rhymes

Caribou

City Of Prague..

Basic Operations

Your Android phone's capability to frustrate you is only as powerful as your fear of the touchscreen and how it works. After you clear that hurdle, understanding how your phone works becomes easier.

Manipulating the touchscreen

The touchscreen works in combination with one or two of your fingers. You can choose which fingers to use, or whether to be adventurous and try using the tip of your nose, but touch the touchscreen you must. Here are some of the many ways you manipulate your phone's touchscreen:

Tap: In this simple operation, you touch the screen. Generally, you're touching an object such as an icon or a control. You might also see the terms *touch* or *press*.

Double-tap: Touch the screen twice in the same location. A double-tap can be used to zoom in on an image or a map, but it can also zoom out. Because of the double-tap's dual nature, I recommend using the pinch and spread operations instead.

Long-press: Tap part of the screen and hold down your finger. Depending on what you're doing, a pop-up menu may appear, or the item you're long-pressing may get "picked up" so that you can drag (move) it around. *Long-press* might also be referred to as *tap and hold.*

Swipe: To swipe, tap your finger on one spot and then move your finger to another spot. Swipes can go up, down, left, or right. This action often moves information on the touchscreen, similar to the way scrolling works on a computer. A swipe can be fast or slow. It's also called a *flick* or *slide.*

Drag: A combination of long-press and then swipe, the drag operation moves items on the screen. Start with the long press, then swipe while keeping your finger on the screen. Lift your finger to complete the action.

Pinch: A pinch involves two fingers, which start out separated and then are brought together. The pinch is used to zoom out on an image or a map. This move may also be called a *pinch close.*

Spread: In the opposite of a pinch, you start with your fingers together and then spread them. The spread is used to zoom in. It's also known as a *pinch open.*

Rotate: Use two fingers to twist around a central point on the touchscreen, which has the effect of rotating an object on the screen. If you have trouble with this operation, pretend that you're turning the dial on a safe.

You can't manipulate the touchscreen while wearing gloves, unless they're gloves specially designed for use with electronic touchscreens, such as the gloves that Batman wears.

Using the navigation icons

Below the touchscreen dwells a series of icons. They can appear as part of the touchscreen itself, or on some phones they may be part of the bezel or even be physical buttons. These are the navigation icons, and they serve specific and uniform functions throughout the Android operating system.

Traditionally you'll find three navigation icons: Back, Home, and Recent. How these icons appear can vary. The most common variations are shown in Table 3-1. This book uses the stock Android Lollipop icons, also shown in the table.

Table 3-1	Navigation Icon Varieties	
Icon	*Lollipop*	*Kit Kat and Earlier*
Home	◯	⌂
Back	◁	↰
Recent	▢	▭

Back: The Back icon serves several purposes, all of which fit neatly under the concept of "back." Tap the icon once to return to a previous page, dismiss an onscreen menu, close a card, hide the onscreen keyboard, and so on.

Home: No matter what you're doing on the phone, touching this icon displays the Home screen. When you're already viewing the Home screen, tap this icon to view the main or center Home screen page.

Recent: Tap the Recent icon to see the Overview, a list of recently opened or currently running apps. Swipe the list to view all the apps. Tap the app's card to switch to that app. To exit from the Overview, tap the Back icon. See the later section "Switching between running apps" for more info on the Overview.

✔ On Android phones without a Recent icon, long-press the Home icon to see the Overview.

✔ Older Samsung phones may sport a Menu icon or button instead of the Recent icon. Tap this icon to display a pop-up menu from which you can choose actions.

✓ The three navigation icons may hide themselves when some apps run. In most cases, the icons are still there but are invisible. Tap the screen to summon the navigation icons. For some full-screen apps and games, swipe the screen from top to bottom to see the icons.

✓ The Back navigation icon may change its orientation, as shown in the margin. Tap this icon to hide the onscreen keyboard, dismiss dictation, or perform other actions similar to using the Back navigation icon.

Setting the volume

The Volume key is found on the side of the phone. Press the top part of the key to raise the volume. Press the bottom part to lower the volume.

As you press the Volume key, a card appears on the touchscreen to illustrate the relative volume level, as shown in Figure 3-1.

Figure 3-1: The volume-controls graphic.

Not every volume card looks like the one shown in Figure 3-1, although they all have a slider. Additional controls let you set specific volumes; tap an icon on the card to view details.

✓ The Volume key controls whatever noise the phone is making when you use it: When you're on a call, the volume controls set the call level. When you're listening to music or watching a video, the volume controls set the media volume.

✓ You can set the volume even when the phone is locked. That means you don't need to unlock the phone to adjust the volume while you're listening to music.

✓ See Chapter 20 for information on setting the volume for specific phone events by using the Settings app.

"Silence your phone!"

How many times have you heard the admonition "Please silence your cell phones"? The quick way to obey this command with your Android phone is to keep pressing the bottom part of the Volume key until the phone vibrates. You're good to go.

✓ Some phones feature a Mute action on the Phone Options card: Press and hold the Power/Lock key and Mute or Vibrate.

✓ You might also find the Sound quick setting. Tap that icon to mute or vibrate the phone. See the later section "Accessing quick settings" for details.

✔ When the phone is silenced or in Vibration mode, an appropriate status icon appears on the status bar. Shown in the margin is what a Vibration mode status icon might look like.

✔ You make the phone noisy again by reversing the directions in this section. Most commonly, pressing the "louder" end of the Volume key restores the phone's sound.

Changing the orientation

Android phones feature a gizmo called an *accelerometer*. It's used by various apps to determine in which direction the phone is pointed or whether you've reoriented the phone from an upright to a horizontal position.

To demonstrate how the phone orients itself, rotate the device to the left or right. Most apps change their orientation to match the new orientation, such as the web browser app, shown in Figure 3-2.

Portrait   Landscape

Figure 3-2: Vertical and horizontal orientations.

The rotation feature may not work for all apps, especially games that present themselves in one orientation only. Also, the Home screen may not reorient itself on all phones.

✔ The onscreen keyboard is more usable when the phone is in its horizontal orientation. See Chapter 4.

✔ Some phones feature a quick setting that locks the orientation. See the later section "Accessing quick settings."

✔ A nifty app for demonstrating the phone's accelerometer is the game *Labyrinth*. You can purchase it at the Google Play Store or download the free version, *Labyrinth Lite*. See Chapter 16 for more information on the Google Play Store.

Home Screen Chores

The Home screen is where you start your Android day. It's the location from which you begin common phone duties, including starting apps. Knowing about the Home screen is an important part of understanding your Android phone.

To view the Home screen at any time, tap the Home navigation icon found at the bottom of the touchscreen. Some phones feature a physical Home button, which performs the same duties as the Home navigation icon.

Exploring the Home screen

A typical Android Home screen is illustrated in Figure 3-3. Several fun and interesting things appear on the Home screen. Find these items on your own phone's Home screen:

Status bar: The top of the Home screen is a thin, informative strip that I call the *status bar*. It contains notification icons and status icons plus the current time. If the status bar disappears, a quick swipe from the top of the screen downward redisplays it.

Notifications: These icons come and go, depending on what happens in your digital life. For example, a new notification icon appears whenever you receive a new email message or have a pending appointment. See the later section "Reviewing notifications."

Phone status: Icons on the right end of the status bar represent the phone's current condition, such as the type of network it's connected to; its signal strength, Wi-Fi status, and battery charge; as well as other items.

App launcher icons: The meat of the meal on the Home screen plate are the app launcher icons. Tapping a launcher icon runs an app.

Status bar

Notifications

Phone Status

Current time

Widget

App icon

Home screen
page index

Favorites tray

Phone app

Apps icon

Folder

Wallpaper

Figure 3-3: The Home screen.

Widgets: These teensy programs can display information as well as let you control the phone, manipulate a feature, access an app, or do something purely amusing.

Folders: Multiple apps can be stored in a folder. Tap the folder to see the app launcher icons inside. Tap a launcher to start that app. See Chapter 19 for more information on folders.

Wallpaper: The background image you see on the Home screen is the wallpaper.

Home screen page index: This series of dots helps you navigate between the various Home screen pages. See the next section.

Favorites tray: The lineup of launcher icons near the bottom of the screen is the favorites tray. It shows the same launchers at the bottom of every Home screen page. Not every phone features a favorites tray.

Phone app: You use the Phone app to make calls. It's kind of a big deal.

Apps icon: Tap this icon to view the collection of apps and widgets available on your phone. See the later section "Finding an app in the Apps drawer."

Ensure that you recognize the names of the various parts of the Home screen. These terms are used throughout this book and in whatever other scant Android phone documentation exists. Directions for using the Home screen gizmos are found throughout this chapter.

- ✔ The Home screen is entirely customizable. You can add and remove icons, add widgets and shortcuts, and even change wallpaper images. See Chapter 20 for more information.

- ✔ Touching part of the Home screen that doesn't feature an icon or a control does nothing — unless you're using the *live wallpaper* feature. In that case, touching the screen changes the wallpaper in some way, depending on the wallpaper that's selected. You can read more about live wallpaper in Chapter 20.

- ✔ You may see numbers affixed to certain Home screen launcher icons. Those numbers indicate pending actions, such as unread email messages, indicated by the icon shown in the margin.

Switching Home screen pages

The Home screen is more than what you see. It's actually an entire street of Home screens, with only one Home screen *page* displayed at a time.

To switch from one page to another, swipe the Home screen left or right. The currently displayed Home screen can be determined by looking at the Home screen page index, which is referenced earlier, in Figure 3-3.

- ✔ On some phones, the main Home screen page is shown by a House icon instead of a dot in the Home screen page index.

- ✔ See Chapter 20 for information on managing home screen pages, adding new pages, and removing pages.

- ✔ The far left Home screen page on some phones is the Google Now app. See Chapter 15 for information on Google Now.

Reviewing notifications

Notifications appear as icons at the top left of the Home screen, as illustrated earlier, in Figure 3-3. To review them, pull down the notifications drawer by dragging your finger from the top of the screen downward. Two variations on the notifications drawer presentation are illustrated in Figure 3-4.

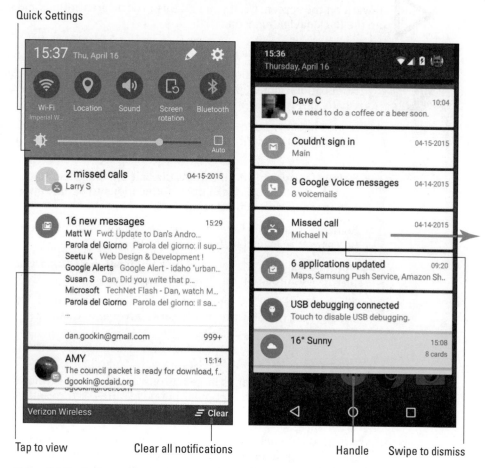

Figure 3-4: Reviewing notifications.

Peruse the notifications by swiping them up and down. To deal with a specific notification, tap it. What happens next depends on the notification, although typically the app that generated the notification runs and shows more details.

After a notification is chosen, it disappears. Or you can dismiss a notification by swiping it to the right or left. To dismiss all notifications, tap the Clear icon, as shown in the margin. If the Clear icon isn't visible, as shown on the right in Figure 3-4, swipe the notifications drawer up or down to locate it.

 To hide the notifications drawer, swipe the notifications drawer handle upward on the screen. Or, if you find this process frustrating (and it can be), tap the Back navigation icon.

✔ Notifications can stack up if you don't deal with them!

✔ When more notifications are present than can be shown, the More Notifications icon appears on the far left end of the status bar, similar to what's shown in the margin.

✔ Some ongoing notifications cannot be dismissed. For example, the USB Connection notification stays active until the USB cable is disconnected.

✔ Swiping away some notifications doesn't prevent them from appearing again in the future. For example, notifications to update apps continue to appear, as do calendar reminders.

✔ Some apps, such as Facebook and Twitter, don't display notifications unless you're logged in. See Chapter 11.

✔ New notifications are heralded by a notification ringtone. Chapter 20 provides information on changing the sound.

✔ Notifications may also appear on the phone's lock screen. Controlling which types of notifications appear is covered in Chapter 21.

Accessing Quick Settings

The Quick Settings appear as large buttons or icons atop the notifications drawer. Use them to access popular phone features or turn settings on or off, such as Bluetooth, Wi-Fi, Airplane Mode, Auto Rotate, and more.

On some phones, such as the Samsung phone shown on the left in Figure 3-4, the Quick Settings appear all the time atop the navigation drawer. Swipe the Quick Settings icons left or right to view the lot. Other phones require that you swipe the notifications drawer downward twice to see quick settings, as illustrated in Figure 3-5.

Dismiss the Quick Actions drawer by touching the Back or Home navigation icons. As long as you have excellent touchscreen skills, you can also banish the Quick Actions drawer by swiping the screen from bottom to top.

✔ Some older Android phones may display the Quick Settings when you swipe the screen from top to bottom, starting at the right end of the status bar.

✔ The Settings icon provides instant access to the Settings app, a popular place to visit as you learn about and configure your Android phone.

Figure 3-5: Quick Settings drawer.

The World of Apps

The Android operating system can pack thrill-a-minute excitement, but it's probably not the only reason you purchased the phone. No, an Android phone's success lies with the apps you obtain. Knowing how to deal with apps is vital to being a successful, happy Android phone user.

Starting an app

To start an app, tap its icon. The app starts.

Apps can be started from the Home screen: Tap a launcher icon to start the associated app. Apps can be started also from the Apps drawer, as described in the later section "Finding an app in the Apps drawer."

You can also start an app found in a Home screen folder: Tap to open the folder.

Quitting an app

Unlike a computer, you don't need to quit apps on your Android phone. To leave an app, tap the Home navigation icon to return to the Home screen. You can keep tapping the Back navigation icon to back out of an app. Or you can tap the Recent navigation icon to switch to another running app.

✔ Occasionally you'll find an app that features a Quit command or Exit command, but for the most part, you don't quit an app on your phone like you quit a program on a computer.

✔ If necessary, the Android operating system shuts down apps you haven't used in a while. You can directly stop apps that have run amok, as described in Chapter 19.

Finding an app in the Apps drawer

The place where you find all apps installed on your Android phone is the *Apps drawer.* Even though you can find launcher icons (app shortcuts) on the Home screen, the Apps drawer is where you need to go to find *everything.*

Wonderful widgets

Like apps, widgets appear on the Home screen. To use a widget, tap it. What happens after that depends on the widget and what it does.

For example, the YouTube widget lets you peruse videos. The Calendar widget shows a preview of your upcoming schedule. A Twitter widget may display recent tweets. Other widgets do interesting things, display useful information, or give you access to the phone's settings or features.

New widgets are obtained from the Google Play Store, just like apps. See Chapter 16 for information. Also see Chapter 19 for details on adding widgets to the Home screen.

To view the Apps drawer, tap the Apps icon on the Home screen. This icon has a different look to it, depending on your Android phone. Figure 3-6 illustrates some variations on the Apps icon, though more varieties may exist.

Figure 3-6: Apps icon varieties.

After you tap the Apps icon, you see the Apps drawer. Swipe through the pages left and right or up and down across the touchscreen.

To run an app, tap its icon. The app starts, taking over the screen and doing whatever magical thing the app does.

- ✔ As you add new apps to your phone, they appear in the Apps drawer. See Chapter 16 for information on adding new apps.

- ✔ Some phones allow you to create folders in the Apps drawer. These folders contain multiple apps, which helps keep things organized. To access apps in the folder, tap the Folder icon.

- ✔ The Apps drawer displays apps alphabetically. On some phones, you can switch to a non-alphabetical viewing grid. With that feature active, it's possible to rearrange the apps in any order you like.

- ✔ For apps that you use all the time, consider creating launcher icons on the Home screen. Chapter 19 describes how.

Switching between running apps

The apps you run on your phone don't quit when you dismiss them from the screen. For the most part, they stay running. To switch between running apps, or any app you've recently opened, tap the Recent navigation icon. You see the Overview, similar to what's shown in Figure 3-7.

To switch to an app, choose it from the list.

- ✔ You can remove an app from the Overview by swiping it off the list. Removing an app from the Overview is pretty much the same thing as quitting an app.

- ✔ For Android phones without a Recent icon, long-press the Home navigation icon to see the Overview.

REMEMBER

✔ The Android operating system may shut down apps that haven't received attention for a while. Don't be surprised when you see an app missing from the Overview. If so, just start it up again as you normally would.

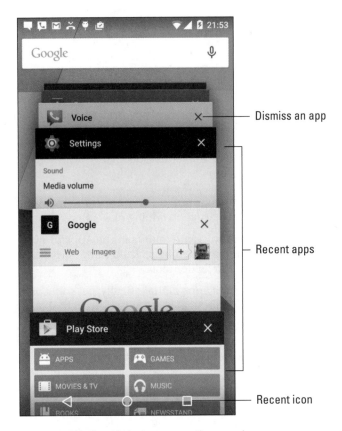

Figure 3-7: The Overview shows recently opened apps.

Common Android Icons

In additional to the navigation icons, various other icons appear while you use your Android phone. These icons serve common functions in your apps as well as in the Android operating system. Table 3-2 lists the most common icons and their functions.

Table 3-2		Common Icons
Icon	*Name*	*Function*
	Action Bar	Displays a pop-up menu. This teensy icon appears in the lower right corner of a button or image, indicating that actions (commands) are attached.
	Add	Adds or creates an item. The plus symbol (+) may be used in combination with other symbols, depending on the app.
	Close	Dismisses a card or clears text from an input field.
	Delete	Removes one or more items from a list or deletes a message.
	Dictation	Lets you use your voice to dictate text.
	Done	Dismisses an action bar, such as the text-editing action bar.
	Edit	Lets you edit an item, add text, or fill in fields.
	Favorite	Flags a favorite item, such as a contact or a web page.
	Overflow	Displays a list of actions, similar to a menu.
	Refresh	Fetches new information or reloads.

(continued)

Table 3-2 (continued)

Icon	Name	Function
	Search	Searches the phone or the Internet for a tidbit of information.
	Settings	Adjusts options for an app.
	Share	Shares information stored on the phone via email, social networking, or other Internet services.

Various sections throughout this book give examples of using the icons. Their images appear in the book's margins where relevant.

- Other common symbols are used on icons in various apps. For example, the standard Play and Pause icons are used in many apps.

- The Overflow might also be referred to as the Action Overflow.

- Samsung phones may display a MORE button instead of the Overflow icon. The MORE button serves the same function.

Text to Type, Text to Edit

4

You may be surprised by the amount of typing you do on an Android phone. Although I seriously doubt that anyone would consider writing the Great American Novel, you instead type text messages, compose email, and take notes as part of your daily phone duties. To help with those tasks, your phone features an onscreen keyboard. It also provides for voice dictation. You can even edit text on the touchscreen just as you would on a word processor. If you find all that stuff overwhelming, you're reading the right chapter.

Onscreen Keyboard Mania

The onscreen keyboard appears on the bottom part of the touchscreen whenever your phone demands text as input. Figure 4-1 illustrates the typical Android keyboard, which is called the Google keyboard. Your phone may use the same keyboard or some variation that looks subtly different. All onscreen keyboards are based on the traditional QWERTY layout and not something designed merely to tick you off.

In Figure 4-1, the onscreen keyboard is shown in alphabetic mode. You see keys from A through Z in lowercase. You also see a Shift key for producing capital letters, and a Delete key, which works to backspace and erase.

Delete

Predictive text   Dictation

Shift   Symbol keyboard   Space key   Return

Figure 4-1: The onscreen keyboard.

The Return key changes its look depending on what you're typing. Your keyboard may show these variations graphically or by labeling the key with text, both of which are shown in Figure 4-2. Here's what each one does:

Return   Search   Go   Next   Done

Figure 4-2: Return key variations.

Return: Just like the Return or Enter key on a computer keyboard, this key ends a paragraph of text. It's used mostly when filling in long stretches of text or when multiline input is needed.

Go: This action key directs the app to proceed with a request, accept input, or perform another action.

Search: You see the Search key when you're searching for something. Tap the key to start the search.

Next: This key appears when you're typing information into multiple fields. Tap this key to switch from one field to the next, such as when typing a username and password.

Done: This key appears whenever you've finished typing text in the final field and are ready to submit input.

The large key at the bottom center of the onscreen keyboard is the Space key. The keys to the left and right may change depending on the context of what you're typing. For example, the / (slash) key or .com key may appear in order to assist in typing a web page or email address. Other keys may change as well, although the basic alphabetic keys remain the same.

- ✐ To display the onscreen keyboard, tap any text field or spot on the screen where typing is permitted.

- ✐ To dismiss the onscreen keyboard, tap the Back icon. This icon may appear as shown in the margin or as a down-pointing chevron.

- ✐ Some onscreen keyboards feature a multifunction key. It may be labeled with the Settings (Gear) icon, a Microphone icon, or another icon. Long-press the multifunction key to view its options.

- ✐ The keyboard reorients itself when you rotate the phone. The onscreen keyboard's horizontal orientation is wide, so you might find it easier to use.

- ✐ I haven't seen one in a while, but some older Android phones featured a physical keyboard. If you'd like to use this type of keyboard with your phone, consider getting a Bluetooth keyboard. The models available for use on tablets also work on Android phones. See Chapter 17 for details on Bluetooth.

The Ol' Hunt-and-Peck

Trust me: No one touch-types on a cell phone. No one. Not even those pre-teens who seem to write text messages at the speed of light. So don't feel bad if you can't type on your Android phone as fast as you can on a computer. On the phone, everything is hunt-and-peck.

Typing one character at a time

The onscreen keyboard is pretty easy to figure out: Tap a letter to produce the character. As you type, the key you touch is highlighted. The phone may give a wee bit of feedback in the form a faint click or vibration.

- ✐ To type in all caps, press the Shift key twice. The Shift key may appear highlighted, the shift symbol may change color, or a colored dot may appear on the key, all of which indicate that Shift Lock is on. Tap the Shift key again to turn off Shift Lock.

- ✐ Above all, it helps to *type slowly* until you get used to the onscreen keyboard.

- ✐ When you make a mistake, tap the Delete key to back up and erase.

- A blinking cursor on the touchscreen shows where new text appears, which is similar to how typing text works on your computer.

- When you type a password, the character you type appears briefly, but for security reasons it's then replaced by a black dot.

- People generally accept the concept that composing text on a phone isn't perfect. Don't sweat it if you make a few mistakes as you type text messages or email, though you should expect some curious replies about unintended typos.

- See the later section "Text Editing," for more details on editing your text.

Accessing special characters

You're not limited to typing only the symbols you see on the alphabetic keyboard. Tap the ?123 key to gain access to additional keyboard layouts, samples of which are shown in Figure 4-3. On some onscreen keyboards, the ?123 key is labeled SYM.

Switch to alpha keyboard

Switch to symbol keyboard

Figure 4-3: Number and symbol keyboards.

In Figure 4-3, the symbol keys are accessed by tapping the key labeled =\<, although some phones label that key with different characters. You might also see page numbers, such as 1/2 or 2/2, to page through multiple pages of characters.

To return to the standard alphabetic keyboard (refer to Figure 4-1), tap the ABC key.

You can access special character keys from the main alphabetic keyboard, providing you know a secret: Long-press a key. When you do, you see a pop-up palette of additional characters, similar to the ones shown for the A key in Figure 4-4.

Drag your finger over a character to select it.

Press and hold.

Figure 4-4: Special symbol pop-up palette thing.

Choose a character from the pop-up palette. If you choose the wrong character, tap the Delete key on the onscreen keyboard to erase the mistyped symbol.

You can use this long-press technique also to access the gray or tiny characters on some variations of the alphanumeric keyboard. Not every phone features those tiny characters.

Typing quickly by using predictive text

As you type, you may see a selection of word suggestions just above the onscreen keyboard. That's the predictive-text feature in action. You can use this feature to greatly accelerate your typing.

In Figure 4-5, I typed the word *I*. The keyboard suggests the words *hope*, *have*, and *will*. Each of those is a logical choice for the next word after *I*. Long-press the center word to view additional choices, as called-out in the figure. Tap a word to insert it into your text.

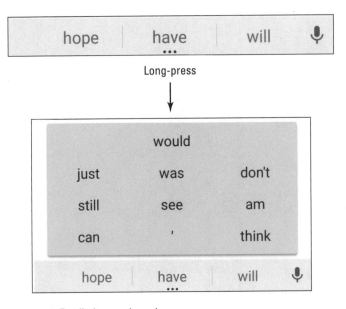

Figure 4-5: Predictive text in action.

When the desired word doesn't appear, continue typing: The predictive text feature makes suggestions based on what you've typed so far. Tap the desired word when it appears.

Predictive text should be active for the Google Keyboard. If not, see Chapter 20 for directions.

Typing without lifting a finger

If you're really after typing speed, consider using gesture typing. It allows you to type words by swiping your finger over the onscreen keyboard, like mad scribbling but with a positive result.

To use gesture typing, drag a finger over letters on the onscreen keyboard. Figure 4-6 illustrates how the word *hello* would be typed in this manner.

Gesture typing isn't used to type a password or email address or for other specific activities. When it doesn't work, type one letter at a time.

See Chapter 20 if you need to activate gesture typing.

Figure 4-6: Using gesture typing to type "hello."

Google Voice Typing

Your Android phone has the amazing capability to interpret your utterances as text. It works almost as well as computer dictation in science fiction movies, though I can't seem to find the command to destroy Alderaan.

✔ The dictation is available whenever you see the Microphone icon. This icon appears on the keyboard as well as other locations, such as search boxes.

✔ On some keyboards, the Microphone icon appears on a multifunction key. Long-press that key to choose its dictation function.

✔ See Chapter 20 for information on enabling dictation, should you have trouble with this feature.

Dictating text

Talking to your phone really works, and works quite well, providing that you tap the Dictation key on the onscreen keyboard and properly dictate your text.

After you tap the Dictation key, a special card appears at the bottom of the screen, similar to the one shown in Figure 4-7. When the text *Tap to Speak* or *Speak Now* appears, dictate your text; speak directly at the phone.

Active (listening) · Paused

Tap to pause · Tap to speak

Show Keyboard

Figure 4-7: Google Voice typing.

As you speak, the microphone icon on the screen flashes. The flashing doesn't mean that the phone is embarrassed by what you're saying. No, the flashing merely indicates that your words are being digested.

The text you utter appears as you speak. To pause, touch the Tap to Pause text on the screen. To use the keyboard, tap the Keyboard icon, as shown earlier, in Figure 4-7.

✒ The first time you try voice input, you might see a description. Tap the OK button to continue.

✒ If you don't like a word that's chosen by the dictation feature, tap the word on the screen. You see a pop-up list of alternatives from which to choose.

✒ You can't use Google Voice typing to edit text. Text editing still takes place on the touchscreen, as described in the later section "Text Editing."

✒ Speak the punctuation in your text. For example, you would say, "I'm sorry comma and it won't happen again" to produce the text *I'm sorry, and it won't happen again* or something close to that.

- Common punctuation you can dictate includes the comma, period, exclamation point, question mark, colon, and new line.

- You cannot dictate capital letters. If you're a stickler for such things, you'll have to go back and edit the text.

- Dictation may not work without an Internet connection.

Uttering s**** words

Your Android phone features a voice censor. It replaces those naughty words you might utter, using the word's first letter followed by an appropriate number of asterisks.

For example, if *spatula* were a blue word and you uttered *spatula* when dictating text, the dictation feature would place s****** on the screen rather than the word *spatula*.

Yeah, I know: silly. Or "s****."

Your Android phone knows a lot of blue terms, including the infamous "Seven Words You Can Never Say on Television," but apparently the terms *crap* and *damn* are fine. Don't ask me how much time I spent researching this topic.

See Chapter 24 if you really want the phone to take naughty dictation.

Text Editing

You probably won't do a lot of text editing on your Android phone. Well, no major editing, such as for a term paper or ransom note. From time to time, however, you may find yourself wanting to fix a word. It's usually a sign that you're over 25; kids no longer seem to care about editing text.

Moving the cursor

The first part of editing text is to move the cursor to the right spot. The *cursor* is that blinking, vertical line where text appears. Then you can type, edit, or paste or simply marvel that you were able to move the cursor hither and thither.

On a computer, you move the cursor by using a pointing device. If your Android phone features a pointing device, such as a thumb ball or arrow keys on a physical keyboard, you can use that device to move the cursor while you edit text. Otherwise, your finger performs the cursor-moving task.

Tap the spot on the text where you want the cursor to appear. To help your accuracy, a cursor tab appears below the text, similar to the one shown in the margin. You can move that tab with your finger to precisely locate the cursor in your text.

After you move the cursor, you can continue to type, use the Delete key to back up and erase, or paste text copied from elsewhere.

You may see a pop-up button by the cursor tab containing a Paste action. Use that button to paste text, as described in the later section "Cutting, copying, and pasting text."

Selecting text

Selecting text on an Android phone works just like selecting text in a word processor: You mark the start and end of a block. That chunk of text appears highlighted on the screen. How you get there, however, can be a mystery — until now!

To select text, long-press a word. Upon success, you see a chunk of text selected, as shown in Figure 4-8.

PASTE

What's a four-letter word for adhesive?

Start block tab   End block tab

Selected text

Figure 4-8: Text is selected.

Drag the start and end markers around the touchscreen to define the block of selected text.

While text is selected, a contextual action bar appears atop the screen, similar to the one shown in Figure 4-9. Your phone may sport a custom action bar with a different look. Either way, you use the action bar to deal with the selected text.

See the later section "Cutting, copying, and pasting text," for information on using the action bar's icons. To delete a selected block of text, tap the Delete key on the onscreen keyboard. To replace the text, type something new.

Close action bar and un-select text   Select all   Copy

Cut   Paste

Figure 4-9: Text-selection contextual action bar.

To cancel text selection, tap the Done icon on the action bar, or just tap anywhere in the text outside the selected block.

- ✔ If long-pressing doesn't select a word, try double-tapping the word.

- ✔ Selecting text on a web page works the same as selecting text in any other app, although I recommend long-pressing and not double-tapping. The difference is that text can only be copied from the web page, not cut or deleted.

- ✔ Seeing the onscreen keyboard is a good indication that you can edit and select text.

- ✔ The action bar's Select All command can be used to mark all text as a single block.

Cutting, copying, and pasting text

Selected text is primed for cutting or copying, which works just like it does in your favorite word processor. After you select the text, choose the proper command from the contextual action bar. To copy the text, choose the Copy command. To cut the text, choose Cut.

Just like on your computer, cut or copied text on your phone is stored in a clipboard. To paste any previously cut or copied text, move the cursor to the spot where you want the text pasted. Tap the Paste button.

- ✔ A quick way to paste text is to look for the Paste button above the cursor tab. To see that button, tap anywhere in the text. Tap the Paste command button to paste in the text.

- ✔ Some phones feature a Clipboard app, which lets you peruse, review, and select previously cut or copied text or images. You might even find the Clipboard icon on the action bar or onscreen keyboard.

- ✔ You can paste text only into locations where text is allowed. Odds are good that if you see the onscreen keyboard, you can paste text.

Dealing with speling errors

Similar to a word processor, your Android phone may highlight misspelled words, presenting them with an intimidating red underline.

To remedy the situation, tap the red-underlined word. You see a pop-up list of alternatives, similar to the one shown in Figure 4-10. Tap a replacement or, if the word is correctly spelled but unknown to the phone, choose to add the word to a personal dictionary.

Pilchuck — Offending word

Pick Huck

Pick hick

Pile hick — Choose alternative

Pill hick

Pic Huck

Add to dictionary

Delete

Figure 4-10: Fixing a misspelled word.

- Words may be autocorrected as you type them. To undo an auto-correction, tap the word again. Choose a replacement word from the predictive-text list, or tap the Replace button to see more options.
- Yes! Your phone has a personal dictionary. See Chapter 24 for details.

Part II
Reach Out and Touch Someone

Keypad	Logs	Favorites	Contacts

Today

(208) 555-3246
Unsaved — 2:19PM

Don Barzini
(212) 555-6797 — 2:06PM

Dan Gookin (2)
(212) 555-2394 — 2:06PM

Don Barzini (6)
(212) 555-6797 — 12:37PM

Don Luciesi (2)
(212) 555-7701 — 9:11AM

Yesterday

Don Luciesi
(212) 555-7701 — 1:33PM

Don Luciesi
(212) 555-7701 — 10:27AM

Joe Biden
(202) 456-1111 — 9:38AM

Voicemail
+11234567890 — 9:31AM

web extras

Explore additional social networking opportunities by visiting
www.dummies.com/extras/androidphones.

In this part . . .

- ☛ Understand how to make phone calls.
- ☛ Explore voicemail and Google Voice.
- ☛ Use the address book.
- ☛ Try text messaging.
- ☛ Send and receive email.
- ☛ Discover the web on a cell phone.
- ☛ Connect with social networking.

5

It's a Telephone

The patent for the telephone was awarded to Alexander Graham Bell in 1876. Telephone-like devices existed before then, and a host of 19th century scientists worked on the concept. Bell beat them all to the patent office, so he gets the credit. And it took many more years for people to patent other, ancillary inventions, including the busy signal (1878), the notion of a second line for teenage girls (1896), and the extension cord, which allowed for simultaneously talking and pacing (1902).

Making a phone call is simple enough; it's the reason you have an Android phone and not an Android tablet. Phone calls aren't the limit of what your device can do. Features such as call waiting, speed dial, call forwarding, and multiple calls are also possible. These features and more are presented in this chapter.

Reach Out and Touch Someone

Until they perfect teleportation, making a phone call is truly the next best thing to being there. It all starts by punching in a number or calling someone from the phone's address book.

Placing a phone call

To place a call on your phone, heed these steps:

1. **Open the Phone app.**

The dialing operation on an Android phone is handled by an app, typically named Phone, although Dialer is also popular. The app's launcher icon is found on the Home screen, often at the bottom in a location called the favorites tray. Despite subtle variations, the Phone app's launcher features a Phone Handset icon, similar to the one shown in the margin.

2. **If necessary, display the dialpad.**

To display the dialpad, either tap the Dialpad icon, as shown in the margin, or a dialpad tab in the Phone app. A typical Dialpad screen is shown in Figure 5-1.

3. **Type a phone number.**

Use the keypad to type the number. You may hear the traditional touchtone sounds as you punch in the number.

If you make a mistake, tap the Delete icon, labeled in Figure 5-1, to back up and erase.

The phone displays matching contacts as you type. Choose a contact to instantly input that person's number. On some phones, choosing a contact dials the number immediately.

4. **Tap the Phone handset (Dial) icon to place the call.**

While a call is active, the screen changes to show contact information, or a contact image when one is available, similar to Figure 5-2.

5. **Place the phone to your ear and wait.**

6. **When the person answers the phone, talk.**

What you say is up to you, though it's good not to just blurt out unexpected news: "I just ran over your cat" or "I think I just saw your daughter on the back of a motorcycle, leaving town at inappropriately high speeds."

Use the phone's Volume key to adjust the volume during the call.

7. **Tap the End Call icon to end the call.**

The phone disconnects. You hear a soft *beep,* which is the phone's signal that the call has ended.

Voicemail

Action Overflow

Matching contacts   Signal strength

Delete

Keypad

Dial

Figure 5-1: The Phone app's dialpad.

 You can do other things while you're making a call: Tap the Home navigation icon to run an app, read old email, check an appointment, or do whatever. Activities such as these don't disconnect you, although your cellular carrier may not allow you to do other things with the phone while you're on a call.

 To return to a call after doing something else, choose the Call in Progress notification icon, similar to the one shown in the margin.

Mute

Speaker

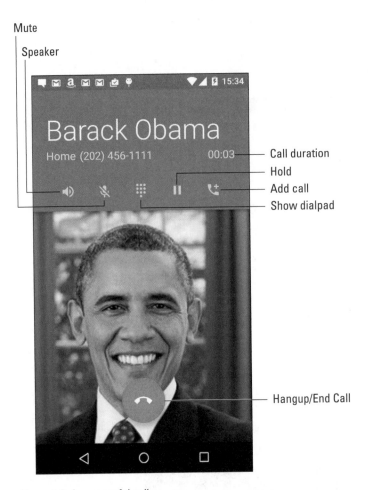

Barack Obama

Home (202) 456-1111   00:03 — Call duration

Hold

Add call

Show dialpad

Hangup/End Call

Figure 5-2: A successful call.

- ✔ Don't worry about the phone's microphone being too far away from your mouth; it picks up your voice just fine.

- ✔ For hands-free operation, use earbuds with a microphone doodle. See Chapter 1 for details on obtaining and using this accessory.

- ✔ You can connect or remove the earphones at any time during a call.

- ✔ You can also go hands-free by using a Bluetooth headset. If a Bluetooth icon doesn't appear on the screen, tap the Speaker icon (refer to Figure 5-2) to ensure that the Bluetooth headset is active. See Chapter 17 for information on Bluetooth.

- ✔ If you're going hands-free, press the phone's Power/Lock key during the call to lock the phone. Locking the phone doesn't disconnect the call.

- ✔ To mute a call, tap the Mute icon, shown earlier in Figure 5-2. A Mute status icon, similar to the one shown in the margin, appears atop the touchscreen.

- ✔ You can't accidentally mute or end a call when the phone is placed against your face; the phone's proximity sensor prevents that from happening.

- ✔ Tap the Speaker icon and hold the phone at a distance to listen and talk, which allows you to let others listen and share in the conversation. The Speaker status icon appears when the speaker is active.

- ✔ Don't hold the phone right at your ear while the speaker is active.

- ✔ If you're wading through one of those nasty voicemail systems, tap the Dialpad icon, labeled in Figure 5-2, so that you can "Press 1 for English" when necessary.

- ✔ See the later section "Multiple Call Mania," for information on using the Add Call and Hold icons.

- ✔ Use the + (Plus) icon below the 0 (zero) key to place an international call. See Chapter 22 for more information.

- ✔ You hear an audio alert whenever the call is dropped or the other party hangs up. The disconnection can be confirmed by looking at the phone, which shows that the call has ended.

- ✔ You cannot place a phone call when the phone has no service; check the signal strength (refer to Figure 5-1). Also see the nearby sidebar, "Signal strength."

- ✔ You cannot place a phone call when the phone is in Airplane mode. See Chapter 22 for information.

Signal strength

One of your Android phone's most important status icons is Signal Strength. It appears in the upper-right corner of the screen, next to the Battery status icon and the time.

The Signal Strength icon features the familiar bars, rising from left to right. The more bars, the better the signal. An extremely low signal is indicated by zero bars. When no signal is available, you may see a red circle with a line through it (the International No symbol) over the bars.

When the phone is out of its service area but still receiving a signal, you see the Roaming icon, which typically includes an *R* near or over the bars. See Chapter 22 for more information on roaming. Also see Chapter 17 for information on the mobile data network status icon.

Dialing a contact

To access your phone's address book, start the Phone app and tap the Contacts tab, which might be titled All Contacts on some phones. Browse the list for someone to call; tap their entry and then tap their phone number or Phone icon to place the call.

You can also dial a contact by first opening the Contacts or People app. The list of contacts in that app is identical to the list accessed by using the Phone app.

- ✔ A special contact category in the phone's address book is Favorites. To quickly access your favorites, tap the Favorites tab in the Phone app.

- ✔ Some variations on the Phone app place the favorite contacts on the main screen or on the Speed Dial screen. See the next section.

- ✔ See Chapter 7 for more information on the phone's address book.

Using speed dial

The speed dial feature isn't available on all Android phones. When it is, you can assign up to eight speed dial entries, which are dialed by long-pressing a number on the Phone app's dialpad. For example, if your bookie is on speed dial 2, long-press the 2 key to instantly dial his number.

To assign a speed dial number, or just review the current settings, heed these steps when using the Phone app:

1. **Display the dialpad.**

2. **Tap the Action Overflow icon.**

    If you don't see the Action Overflow icon on the screen, tap the More button.

3. **Choose Speed Dial or Speed Dial Setup.**

    If you don't see that these are similar actions, the Phone app most likely lacks a speed dial feature.

    Most carriers configure number 1 as the voicemail system's number. The remaining numbers, 2 through 9, are available to program.

4. **Tap an item on the list, or tap the Add icon.**

    The item may say Add Speed Dial or Not Assigned as opposed to being blank.

5. **Choose a contact.**

6. **Repeat Steps 4 and 5 to assign more speed dial contacts.**

When you're done adding numbers, tap the Back navigation icon to exit the speed dial screen.

To remove a speed dial entry, long-press it and choose the Delete or Remove action. Or, in some cases, tap the Minus (Remove) icon to the right of the speed dial entry.

Adding pauses when dialing a number

When you tap the Phone icon to dial a number, the number is instantly spewed into the phone system, like water out of a hose. If you need to pause the number as it's dialed, you need to know how to insert secret pause characters. Two are available:

- ✔ The comma (,) adds a 2-second pause.
- ✔ The semicolon (;) adds a wait prompt.

To insert the pause or wait characters into a phone number, obey these directions:

1. **Type the number to dial.**

2. **At the point that the pause or wait character is needed, tap the Action Overflow icon.**

    On some phones, tap the More button.

3. **Choose the action Add 2-Sec Pause or Add Wait.**

4. **Continue composing the rest of the phone number.**

When the number is dialed and the comma (,) is encountered, the phone pauses two seconds and then dials the rest of the number. When the semicolon (;) is encountered, the phone prompts you to continue. Tap the Yes or OK button to continue dialing the rest of the number.

- ✔ The comma (,) and semicolon (;) can also be inserted into the phone numbers you assign to contacts in the phone's address book. See Chapter 7.

- ✔ Alas, you cannot program an interactive phone number, such as one that pauses and lets you provide input and then continues to dial. You have to perform that task manually on an Android phone.

It's For You!

Who doesn't enjoy getting a phone call? It's an event! Never mind that it's the company that keeps calling you about lowering the interest rate on your credit cards. The point is that someone cares enough to call. Truly, your Android phone ringing can be good news, bad news, or mediocre news, but it always provides a little drama to spice up an otherwise mundane day.

Receiving a call

Several things can happen when you receive a phone call on your Android phone:

- ✔ The phone rings or makes a noise signaling you to an incoming call.
- ✔ The phone vibrates.
- ✔ The touchscreen reveals information about the call, as shown at the top of Figure 5-3.
- ✔ The car in front of you explodes in a loud fireball as your passenger screams something inappropriately funny.

Figure 5-3: You have an incoming call.

That last item happens only in Bruce Willis movies. The other three possibilities, or a combination thereof, are your signals that you have an incoming call.

Work the touchscreen to answer the call. The stock Android method is shown in Figure 5-3: Swipe the Answer icon to the left to answer. Other methods for answering a call are shown in Figure 5-4. Generally, you swipe the Answer icon in a specific direction.

The center image in Figure 5-4 shows how a call may look while you're using the phone: Tap the Answer button to take the call.

When you choose to answer the call, place the phone to your ear. If you're using earbuds or a Bluetooth headset, tap the button to answer. Say "Hello," or, if you're in a grouchy mood, say "What?" loudly.

Figure 5-4: Answering a call.

Tap the End Call button to hang up when you're done with the call. If the other party hangs up first, the call ends automatically.

- You don't have to work a screen lock to answer a call; simply slide the Answer icon. If you want to do other things while you're on the call, you have to work the screen lock. See Chapter 21 for information on screen locks.

- The contact's picture appears only when you've assigned a picture to the contact. Otherwise, a generic contact image shows up (refer to the far right side of Figure 5-4).

- The sound you hear when the phone rings is known as the *ringtone.* You can configure your phone to play a number of ringtones, depending on who is calling, or you can set a universal ringtone. Ringtones are covered in Chapter 20.

Rejecting a call

Several options are available when you don't want to receive an incoming call.

Let the phone ring: Just do something else or pretend that you're dead. You can silence the ringer by pressing the phone's Volume key.

Dismiss the call: Slide the Answer icon to the Dismiss icon, or just swipe the Decline icon, as illustrated earlier, in Figure 5-4.

Reply with a text message: Choose the text message rejection option, which sends the caller a text message but doesn't pick up the line.

In all cases, when you don't answer the phone, the call is sent to voicemail. See Chapter 6 for information on voicemail. Chapter 6 also covers the call log, which shows a list of recent calls, incoming, missed, and rejected.

Rejecting a call with a text message

Rather than just dismiss a call, you can reply to the call with a text message. Choose the Text Message icon or drawer when dismissing a call, shown in Figures 5-3 and Figure 5-4, respectively. Some phones may not display the text message rejection option until after you dismiss the call.

After choosing the text message rejection option, select a text message. Figure 5-5 shows how such a screen may look.

After you choose a text message, the incoming call is dismissed. In a few cellular seconds, the person who called receives the message.

- Not every phone offers the text message rejection feature.

- The method for adding, removing, or editing the call rejection messages differs from phone to phone. Generally, tap the Action Overflow icon or More button while using the Phone app. Choose Settings. Look for a Quick Responses or Call Rejection item.

- See Chapter 8 for more information on text messages.

REJECT CALL WITH MESSAGE
+ Compose new message
I'm driving and shouldn't be using my phone. Heck, I'm risking death by...
I'm in a meeting. Please call later. Thanks
Can't talk now, but I can text. What's up?

Figure 5-5: Text message rejection selection.

Multiple Call Mania

As a human being, your brain limits your ability to hold more than one conversation at a time. Your phone's brain, however, lacks such a limitation. It's entirely possible for an Android phone to handle more than one call at a time.

Putting someone on hold

It's easy to place a call on hold using your Android phone — as long as your cellular provider hasn't disabled that feature. Simply tap the Hold icon, shown in the margin. Refer to Figure 5-2 for the Hold icon's location on the call-in-progress screen.

To take the call out of hold, tap the Hold icon again. The icon may change its look, for example, from a Pause symbol to a Play symbol.

Fret not if your phone's call-in-progress screen lacks the Hold icon. Rather than hold the call, mute it: Tap the Mute icon. That way, you can sneeze, scream at the wall, or flush the toilet and the other person will never know.

Receiving a new call when you're on the phone

You're on the phone, chatting it up. Suddenly, someone else calls you. What happens next?

Your phone alerts you to the new call, perhaps by vibrating or making a sound. Look at the touchscreen to see who's calling and determine what to do next. You have three options:

- ✔ **Answer the call.** Slide the Answer icon just as you would answer any incoming call. The call you're on is placed on hold.

- ✔ **Send the call directly to voicemail.** Dismiss the call as you would any incoming phone call.

- ✔ **Do nothing.** The call eventually goes into voicemail.

When you select to answer the second call, additional options become available to manage both calls. Use special icons on the call-in-progress screen to perform special, multi-call tricks:

Swap/Switch Calls: To switch between callers, tap the Swap or Switch Calls icon on the touchscreen. Each time you tap this icon, the conversation switches to the other caller, and the current person is put on hold.

Merge Calls: To combine all calls so that everyone is talking (three people total), tap the Merge Calls icon. This icon may not be available if the merge feature is suppressed by your cellular provider.

End Call: To end a call, tap the End Call icon, just as you normally do. You're switched back to the other caller.

To end the final call, tap the End Call or Hang Up icon, just as you normally would.

- ✔ The number of different calls your phone can handle depends on your carrier. For most subscribers in the United States, your phone can handle only two calls at a time. In that case, a third person who calls you either hears a busy signal or is sent directly into voicemail.

- ✔ If the person on hold hangs up, you may hear a sound or feel the phone vibrate when the call is dropped.

- ✔ When you end a second call on the Verizon network, both calls may appear to have been disconnected. That's not the case: In a few moments, the call you didn't disconnect "rings" as though the person is calling you back. No one is calling you back, though: The phone is simply returning you to that ongoing conversation.

Making a conference call

You can call two different people by using your Android phone's merge calls feature, also known as a conference call. Start by connecting with the first person and then *add* a second call. Soon, everyone is talking. Here's how it works:

1. **Phone the first person.**

2. **After the call connects and you complete a few pleasantries, tap the Add Call icon.**

   The Add Call icon may appear as shown in the margin. If not, look for the Plus (+) icon. After you tap that icon, the first person is placed on hold.

3. **Dial the second person.**

   You can use the dialpad or choose the second person from the phone's address book or the call history.

   Say your pleasantries and inform the party that the call is about to be merged.

4. **Tap the Merge or Merge Calls icon.**

   The two calls are now joined: Everyone you've dialed can talk to and hear everyone else.

5. **Tap the End Call icon to end the conference call.**

   All calls are disconnected.

When several people are in a room and want to participate in a call, you can always put the phone in Speaker mode: Tap the Speaker icon on the ongoing call screen.

Your Android phone may feature the Manage icon while you're in a conference call. Tap this icon to list the various calls, to mute one, or to select a call to disconnect.

Forward Calls Elsewhere

You have a choice for dealing with incoming calls. You can ignore them, silence the ringer, banish them to email, or admit defeat and answer. You can also choose to automate the process by forwarding one call, some calls, or all incoming calls.

Forwarding phone calls

Call forwarding is the process by which you reroute an incoming call. For example, you can send all your calls to the office while you're on vacation.

Then you have the luxury of having your cell phone and still making calls but freely ignoring anyone who calls you.

The options for call forwarding on your phone can be set by using either the Android operating system itself or the controls set up by your cellular provider.

To confirm that call-forwarding options can be set by using the Android operating system, follow these steps:

1. **Open the Phone app.**

2. **Tap the Action Overflow icon.**

3. **Choose Settings or Call Settings.**

4. **Choose Call Forwarding.**

   This item might be found by first choosing the Calls item.

5. **Choose a call forwarding option.**

   Sometimes only one option is available: a phone number to use for all incoming calls. You might instead see separate options, such as these:

   - *Always Forward:* All incoming calls are sent to the number you specify; your phone doesn't even ring. This option overrides all other forwarding options.

   - *When Busy:* Calls are forwarded when you're on the phone and choose not to answer. This option normally sends a missed call to voicemail.

   - *When Unanswered:* Calls are forwarded when you choose not to answer the phone. Normally, the call is forwarded to voicemail.

   - *When Unreached:* Calls are forwarded when the phone is turned off, out of range, or in Airplane mode. As with the two previous settings, this option normally forwards calls to voicemail.

6. **Set the forwarding number.**

   Or you can edit the number that already appears. For example, you can type your home number for the Forward When Unreached option so that your cell calls are redirected to your home number when you're out of range.

   To disable the option, tap the Turn Off button.

7. **Tap the Back navigation icon when you're done making settings.**

If your phone lacks Call Forwarding settings, you must rely upon the cellular carrier to set up and forward your calls. For example, Verizon in the United States uses the call forwarding options described in Table 5-1.

Table 5-1	Verizon Call-Forwarding Commands	
To Do This	*Input First Number*	*Input Second Number*
Forward unanswered incoming calls	*71	Forwarding number
Forward all incoming calls	*72	Forwarding number
Cancel call forwarding	*73	None

So, to forward all calls to (714) 555-4565, open the Phone app and dial the number ***727145554565**. You hear only a brief tone after dialing, and then the call ends. After that, any call coming into your phone rings at the other number.

You must disable call forwarding to return to normal cell phone operations: Dial ***73.**

Also: Call forwarding may affect your phone's voicemail service. See Chapter 6 for details.

Sending a contact directly to voicemail

You can configure your phone to forward any of your phone's address book contacts directly to voicemail. It's a great way to deal with a pest! Follow these steps:

1. **Open the phone's address book app.**

   The app is called Contacts. See Chapter 7 for details.

2. **Choose a contact, someone who's bothering you.**

   Display that person's contact information on the touchscreen.

3. **Edit the contact: Tap the Edit icon.**

   On some phones, instead of editing the contact, tap the More button.

4. **Tap the Action Overflow icon and choose All Calls to Voicemail.**

   The command might instead be titled Add to Reject List.

When a contact has been banished, you'll never receive another call from that person. Instead, you'll see a voicemail notification, but only when the caller choses to leave you a message.

This feature is one reason you might want to retain contact information for someone with whom you never want to have contact.

6

Missed Calls and Voicemail

*T*he genius behind the mobile phone is that you can receive a call anywhere. I'm not implying that you'll never miss a call. In fact, some calls you'll miss on purpose: You whip out the phone, check the touchscreen to see who's calling, frown, and then dismiss the call, banishing it to voicemail. That happens a lot.

As a type of a computer, your Android phone dutifully keeps track of all calls, not only missed, but those you place and answer as well. Missed calls aren't sentenced to the digital ether, either: They can visit a place called voicemail. And that voicemail is more than just a "Leave your tone at the message" feature on an Android phone.

Voicemail 📞 ✎

Marie Antoinette
(211) 555-1793
Received: 14:12 (9 secs)

Hi Dan, This is Mary. Please call me back. I r
someone to talk to bye.

Who Called Who When?

All phone calls made on your Android phone are noted: To whom, date, time, duration, and whether the call was incoming, outgoing, missed, or dismissed. I call this feature the *call log,* although on your phone it may be referenced as Recent Calls or Call History. The place to look is in the Phone app.

Dealing with a missed call

When you miss a call, a Missed Call notification icon appears atop the touch-screen. This icon, similar to the one in the margin, specifically applies to a call you didn't pick up; it doesn't appear for calls you dismiss or calls missed because the phone was turned off or in Airplane mode.

To deal with the notification, display the notification drawer. When the notification represents a single missed call, you see details plus a few buttons, similar to what's shown in Figure 6-1. Tap the Call Back button to immediately return the call.

Return call   Send text   View call log

Figure 6-1: A missed-call notification.

When several calls have been missed, the missed-call notification shows a list of contacts or phone numbers. Tap that notification to view the call history, where you can return missed calls. See the next section.

✔ Not every Android phone shows the missed-call notification as it appears in Figure 6-1.

✔ The phone doesn't consider a call you've dismissed as being missed.

✔ Refer to Chapter 3 for specifics on accessing the notification drawer.

Reviewing the call log

Your Android phone tracks all calls in a location I refer to as the call log. It's found in the Phone app: Tap the Recents tab, which might be titled Recent, History, or Logs. Figure 6-2 shows what a call log might look like.

Ignored call

Missed call   Number of calls

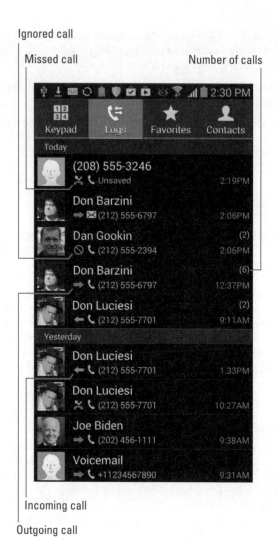

Incoming call

Outgoing call

Figure 6-2: The call log.

Swipe through the list to examine recent calls. The list is sorted so that the most recent calls appear at the top. Information appears associated with each call, such as the date and time, call duration, and whether the call was incoming, outgoing, missed, or ignored.

To see more details about the call, choose its item in the list. You can view more details, see a call history specific to that number or contact, or return the call.

✔ Using the call log is a quick way to add a recent caller as a contact: Tap an item in the list, and choose a Create Contact or Add to Contacts item. See Chapter 7 for more information about contacts.

✔ Remove an item from the call log by long-pressing its entry. Tap the Delete (Trash) icon atop the screen.

✔ To clear the call log, tap the Action Overflow icon and choose the Clear Call History or similar action. Some phones may require you to long-press an entry and then select all entries to clear the call log.

Boring Carrier Voicemail

The most basic, and most stupid, form of voicemail is the free voicemail service provided by your cell phone company. This standard feature has few frills and nothing that stands out differently for your nifty Android phone.

Carrier voicemail picks up missed calls as well as calls you thrust into voicemail. A notification icon, looking similar to the one shown in the margin, appears whenever someone leaves you a voicemail message. You can use the notification to dial into your carrier's voicemail system, listen to your calls, and use the phone's dialpad to delete messages or repeat messages or use other features you probably don't know about because no one ever pays attention to them.

Setting up carrier voicemail

If you haven't yet done it, set up voicemail on your phone. I recommend doing so even if you plan on using another voicemail service, such as Google Voice. That's because carrier voicemail, despite all my grousing, remains a valid and worthy fallback option when those other services don't work.

Even if you believe your voicemail to be set up and configured, consider churning through these steps, just to be sure:

1. **Open the Phone app.**

2. **Tap the Action Overflow icon.**

   On some phones, tap the More button.

3. **Choose Settings.**

4. **Choose Call Settings.**

   On some phones, this item is titled Calls. Some phones may skip this step, showing a Voicemail item on the main Settings screen.

5. **Choose Voicemail or Voicemail Service.**

6. **Choose Service to determine which voicemail service the phone uses.**

   The options are either Carrier (or Your Carrier) and Google Voice, although only one option might be available.

When the Carrier option is chosen, the phone number used for voicemail is the carrier's voicemail service. You can confirm that number by choosing the Setup or Voicemail Settings item on the Voicemail screen.

After performing the steps in this section, call the carrier voicemail service to finish the setup: Long-press the 1 key in the Phone app to dial voicemail. Set your name, a voicemail password, a greeting, and various other steps as guided by the cellular provider's cheerful robot.

> ✔ Some versions of the Phone app feature a Voicemail icon on the dialpad. Use it to quickly connect to your carrier's voicemail service.

> ✔ Your phone may also feature a Voicemail app, which you can use to collect and review your messages.

> ✔ Complete the voicemail setup by creating a customized greeting. If you don't, you may not receive voicemail messages, or people may believe that they've dialed the wrong number.

Retrieving your messages

When you have a voicemail message looming, a notification icon appears on the status bar, similar to the one shown in the margin. You can either pull down this notification to connect to the voicemail service or dial into the voicemail service by long-pressing the 1 key on the Phone app's dialpad.

What happens next depends on how your carrier has configured its voicemail service. Typically, you have to input your PIN or password, and then new messages play or you hear a menu of options. My advice: Look at the phone so that you can see the dialpad, and tap the Speaker icon so that you can hear the prompts.

To help you remember the prompts, write them down here:

Press _____ to listen to the first message.

Press _____ to delete the message.

Press _____ to skip a message.

Press _____ to hear the menu options.

Press _____ to hang up.

While you're at it, in case you ever forget your voicemail password, write it down here:

_____

The Wonders of Google Voice

Perhaps the best option I've found for working with voicemail is something called Google Voice. It's more than just a voicemail system: You can use *Google Voice* to make phone calls in the United States, place cheap international calls, and perform other amazing feats. In this section, I extol the virtues of using Google Voice as the voicemail system on your Android phone.

- ✔ Even when you choose to use Google Voice, I still recommend setting up and configuring the boring carrier voicemail, as covered earlier in this chapter.

- ✔ With Google Voice configured as your phone's voicemail service, your voicemail messages arrive in the form of a Gmail message. The message's text is a transcription of the voicemail message.

- ✔ Better than reading a Gmail message is receiving a Google Voice message by using the Google Voice app. See the later section "Using the Google Voice app."

- ✔ You may need to reset Google Voice after using call forwarding. That's because Google Voice relies upon forwarding unanswered or dismissed calls. See Chapter 5 for more information on call forwarding.

Configuring Google Voice

To configure your Android phone for use with Google Voice, you must first create a Google Voice account. Start your adventure by visiting the Google Voice home page at `https://voice.google.com`. I recommend using a computer to complete these steps.

Follow the directions on the screen. Log in to your Google account, if necessary, and agree to the terms of service.

You're prompted to input information about the phone, and Google calls your cell phone to confirm the number and verify the service.

Eventually you're given directions on how to forward missed calls to the Google Voice service. When the Google Voice access number appears on the computer screen, write it down for future reference:

My Google Voice access number: (_____) _____ - _____

I also recommend creating a contact for your Google Voice access number. See Chapter 7 for information on creating new contacts. And while you're at it, consider bookmarking the Google Voice website on your computer.

Although this step configures a Google Voice account, you still need to program your phone to use Google Voice as its voicemail service. That's best accomplished by installing the Google Voice app. See the next section.

Using the Google Voice app

Google Voice transcribes your voicemail messages, turning the audio from the voicemail into a text message you can read. The messages all show up eventually in your Gmail inbox, just as though someone sent you an e-mail rather than left you voicemail. It's a good way to deal with your messages, but not the best way.

The best way to handle Google Voice is to use the Google Voice app, available from the Google Play Store. (See Chapter 16 for details on obtaining apps at the Google Play Store.)

After the Google Voice app is installed, you have to work through the setup, which isn't difficult: The goal is to switch over the phone's voicemail number from the carrier's voicemail system to Google Voice. Eventually, you see the app's main interface, which looks and works similarly to an email program. You can review your messages or tap a message to read or play it, as illustrated in Figure 6-3.

When new Google Voice messages come in, you see the Google Voice notification icon, as shown in the margin. To read or listen to the message, pull down the notifications and choose the item labeled Voicemail from *whomever*.

✏ With Google Voice installed, you see two notices for every voicemail message: one from Google Voice and another for the Gmail message.

✏ The Google Voice app works only after you acquire a Google Voice account and add your Android phone's number to that account. See the earlier section "Configuring Google Voice."

✏ The text *Transcript Not Available* appears whenever Google Voice is unable to create a text message from your voicemail or whenever the Google Voice service is temporarily unavailable.

Contact info (if available)   Incoming phone number

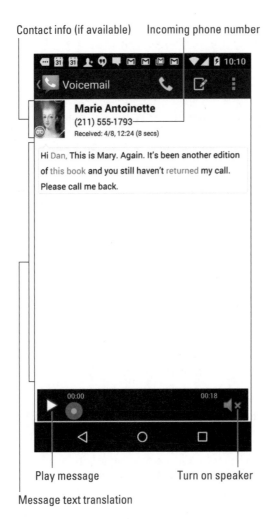

Play message   Turn on speaker

Message text translation

Figure 6-3: Voicemail with the Google Voice app.

The Address Book

In This Chapter

▶ Examining your phone's contacts

▶ Sorting and searching contacts

▶ Adding a new contact

▶ Editing contact information

▶ Putting a picture on a contact

▶ Working with favorites

▶ Deleting contacts

*O*nce upon a time, humans went to the trouble of memorizing phone numbers. They didn't commit everyone's number to memory — just a few key contacts. So if a kid were stranded at the bowling alley, he could phone Mom's workplace, Grandpa, or even a neighbor lady to ask for a lift.

Today, it's your phone that memorizes phone numbers — and more. That's the duty of the phone's address book app, which also stores email addresses, physical addresses, and even random information such as birthdays and anniversaries. It's indispensable.

PHONE

202-456-1111

?ME

The People You Know

The phone's address book is invaluable. It's used by other apps, such as Gmail, Email, Hangouts, and (most importantly) the Phone app. As a bonus, the address book is probably full of people already; your Gmail contacts are instantly synchronized with the address book app, as are social networking contacts and any contacts associated with other accounts you've added to the phone.

Accessing the address book

The stock Android name for the phone's address book app is Contacts. It might also be called People on some phones. Look for the app's launcher icon on the Home screen or in the Apps drawer, as described in Chapter 3.

You can also access the list of contacts from within the Phone app: Tap the Contacts tab to view the list.

Figure 7-1 shows how the Contacts app might look. Its appearance on your phone may be different.

Index

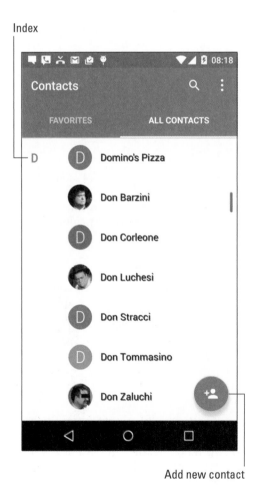

Add new contact

Figure 7-1: The Contacts list.

Scroll the list by swiping your finger on the touchscreen. You can use the index on the side of the screen (refer to Figure 7-1) to quickly scroll the list up and down. Large letters may appear as you scroll, to help you find your place.

When you locate a specific contact, tap the entry in the list. You see more information, similar to what's shown in Figure 7-2.

Edit contact

Favorite

Text message

Send email

Call phone

Figure 7-2: More details about a contact.

The list of activities you can do with the contact depends on the information shown and the apps installed on the phone. Here are some options:

Place a phone call. To call the contact, tap one of the contact's phone entries. More than one might be shown, such as Home or Mobile.

Send a text message. Tap the Text Message icon (refer to Figure 7-2) to open the phone's text messaging app and send the contact a message. See Chapter 8 for information about text messaging.

Send email: Tap the contact's email address to compose an email message. Chapter 9 covers using email on your phone.

Locate the contact's address. When the contact's information shows a home or business address, tap that item to summon the Maps app and view the location. Refer to Chapter 12 to see all the fun stuff you can do with Maps.

Some tidbits of information that show up for a contact don't have an associated action. For example, the phone won't sing "Happy Birthday" when you tap a contact's birthday information.

✓ Tap the Back icon to view the main address book list after viewing a contact's information.

✓ Not every contact has a picture, and the picture can come from a number of sources (Gmail or Facebook, for example). See the later section "Adding a contact picture" for more information.

✓ Many Android phones feature an account named Me. That's your own information as known by the phone. The Me account may be in addition to your other accounts shown in the address book.

✓ Some cellular providers add accounts to your phone's address book. You may see entries such as BAL, MIN, or Warranty Center. Those aren't real people; they're shortcuts to various services. For example, the BAL contact is used on Verizon phones to get a text message listing your current account balance.

✓ Also see the later section "Joining identical contacts" for information on how to deal with duplicate entries for the same person.

Sorting the address book

The phone's address book displays contacts in a certain order, such as alphabetically by first name. If that presentation drives you nuts, change the order to whatever you like by following these steps:

1. **Tap the Action Overflow icon as shown in the margin and choose Settings.**

On some Samsung phones, tap the MORE button, and then choose Settings.

2. **Choose Sort By.**

3. **Choose First Name or Last Name, depending on how you want the Contacts list sorted.**

   I prefer contacts sorted by first name, which is how the Contacts app normally does it.

4. **Choose Name Format.**

5. **Choose First Name First or Last Name First.**

   I prefer First Name First, which again is how the Contacts app presents the list.

The address book is updated, displayed per your preferences.

Searching contacts

Rather than scroll your phone's address book with angst-riddled desperation, follow these steps:

1. **Tap the Search icon.**

   The Find Contacts text box appears.

2. **Type a few letters of the contact's name.**

   As you type, the list of contacts narrows to match the search text.

3. **Tap a matching name to view the contact's information.**

To clear a search, tap the X at the right side of the search text box. To exit the search screen, tap the Back navigation icon.

Make New Friends

Having friends is great. Having more friends is better. You'll find myriad ways to add new friends and place new contacts into your phone. This section lists a few of the more popular and useful methods.

Creating a new contact from scratch

Sometimes it's necessary to create a contact when you meet another human being in the real world. In that case, you have more information to input, and it starts like this:

1. **Tap the Add New Contact icon in the Contacts app.**

   Refer to Figure 7-1 for the icon's look and location, although it may appear as a plus-sign icon at the top of the screen.

2. **Choose an account with which to associate the contact.**

   Account choices may be presented directly, or you may have to choose an account from the action bar.

   I recommend choosing Google, unless you use another account, such as Yahoo!, listed as your main Internet mail account.

   Do not choose the Device or Phone account. When you do, the contact information is saved only on your Android phone. It isn't synchronized with the Internet or any other Android devices.

3. **Fill in the contact's information.**

   Fill in the text fields with the information you know. The more information you provide, the better.

   Some items feature an action bar from which you can choose options, such as whether a phone number is Work, Home, or Mobile.

   To add a second phone number, email address, or location, tap the chevron or plus-sign icon next to an item.

   The Contacts app might sport the Add Another Field button. Use that button whenever you want to add more details for the contact, such as a birthday or website.

4. **Tap the Done or Save button to complete the editing and create the new contact.**

The new contact is created. As a bonus, it's also automatically synced with your Google account, or with whichever account you chose in Step 3.

Adding a contact from the call log

A quick and easy way to build up your phone's address book is to add people as they phone you. To do so, check the call log:

1. **Open the Phone app.**

2. **Display the call log.**

   Tap the Recents, Log, or History tab. See Chapter 6 for specific directions.

   Unknown phone numbers appear by themselves, without a contact picture, name, or other details.

3. **Display details about the phone number for an incoming call.**

   If the details aren't presented right away, long-press the entry or tap the Details button.

After displaying details about the unknown number you have two options: Add it to an existing contact or create a new contact. Use either option to add a contact, although not every Android phone may present these actions.

To add the unknown number to an existing contact, choose the Update Existing option. Locate the contact in the address book list, and the phone number is automatically associated with that person or business.

Creating a new contact works just like creating a contact from scratch, with the exception that the phone number is already entered. Fill in other information: name, email, and what you know (though all that's really needed is the name). Save the new contact.

Creating a contact from an email message

Perhaps one of the easiest ways to build up the phone's address book is to create a contact from an email message. Follow these general steps when you receive a message from someone not already in the phone's address book:

1. **View the email message.**

   You can't add a contact from the inbox; tap the message to view its contents.

2. **Tap the icon by the sender's name or tap the name itself.**

   The Gmail app uses a letter icon for unknown contacts, such as the H shown in the margin. The Email app may show a generic human icon, in which case tap that icon. Otherwise, tap the name itself.

3. **Tap the OK button to create the new contact. Or, if prompted, choose the option to create a new contact.**

   You might also see an option to update an existing contact. If so, choose that contact from the list.

4. **Add information about the new contact.**

   Refer to specific directions earlier in this chapter. The contact's email address is provided automatically. For updating an existing contact, the email address is added to their information.

5. **Save or update the contact.**

   Tap the Save or Done button.

If you create a new contact, double-check the name. It might be added improperly, using the contact's email address or a name you don't normally associate with that contact.

Importing contacts from a computer

Your computer's email program is doubtless a useful repository of contacts you've built up over the years. You can export these contacts from your computer's email program into your Android phone. It's not the easiest thing to do, but it's possible.

The key is to save or export your computer email program's records in the vCard (vcf) file format. These records can be read by the phone's address book app. The method for exporting contacts varies, depending on the email program:

- **In the Windows Live Mail program,** choose Go⇨Contacts, and then choose File⇨Export⇨Business Card (.VCF) to export the contacts.

- **For Microsoft Outlook,** you don't need to do a thing. Outlook contacts are automatically synchronized when you add your Exchange Server account to the phone. Refer to Chapter 2.

- **In Windows Mail,** choose File⇨Export⇨Windows Contacts, and then choose vCards (Folder of .VCF Files) from the Export Windows Contacts dialog box. Click the Export button.

- **On the Mac,** open the Address Book program and choose All Contacts. Choose File⇨Export⇨Export vCard to save the vCards as a single file.

After the vCard files are created, transfer them to your phone. The most effective way is to connect your phone to a computer and use a USB cable to move over the files. See Chapter 18 for specifics.

With the vCard files stored on the phone, follow these steps in the Contacts app to complete the process:

1. **Tap the Action Overflow icon.**

2. **Choose Import/Export.**

   If you don't see this action, first choose Settings and then Import/Export.

3. **Choose Import from Storage.**

   This command may be titled differently, but the word *import* is in there somewhere.

4. **Choose to save the contacts to your Google account.**

5. **If prompted, choose the option Import All vCard Files.**

6. **Tap the OK button.**

   The imported contacts are synchronized to your Google account, or to whichever account you choose (Step 4), which instantly creates a backup copy.

The importing process may create some duplicates. That's okay: You can join two entries for the same person in your phone's address book. See the section "Joining identical contacts," later in this chapter.

Address Book Management

Sure, some folks just can't leave well enough alone. For example, some of your friends may change their phone numbers. They may move. They may finally get rid of their 20-year-old AOL email addresses. When such things occur, you must undertake the task of address book management.

Making basic changes

To make minor touch-ups on any contact, start by locating and displaying the contact's information in the address book app. Tap the Edit icon, which may look similar to the Pencil icon shown in the margin. If this icon isn't available or obvious, tap the Action Overflow icon or Menu button and choose the Edit command.

Tap a field to edit information. Tap the Add icon or chevron next to an existing item to add another item, such as a second phone number or email address.

Some contact information cannot be edited. For example, fields pulled in from social networking sites can be edited only by that account holder on the social networking site.

When you're done editing, tap the Save or Done button.

Adding a contact picture

Spice up your phone's address book by assigning images to contacts. You can use an actual picture of the contact, a photo that reminds you of the contact, or something wholly inappropriate.

To use the phone's camera to set a contact image, follow these steps:

1. **Edit the contact's information.**

2. **Tap the contact's picture or the Change button that appears by the picture.**

   If nothing happens after tapping the picture, tap the Edit icon, and then tap the picture.

3. **Choose the Take New Photo action.**

   The action might be titled Take Picture or something similar. On some phones, you may have to tap the X icon to first remove the photo before you can take a new one.

4. **Use the phone's camera to snap a picture.**

   Chapter 13 covers using the Camera app. Tap the Shutter icon to take the picture.

5. **Review the picture.**

   Not every phone lets you review the image — which is a good thing. If prompted, confirm that the image is up to snuff. If not, tap the Retry icon. Refer to Table 7-1 for definitions of the picture confirmation icons.

6. **Tap the Done icon to confirm the new image and prepare for cropping.**

7. **Crop the image, as illustrated in Figure 7-3.**

   Adjust the cropping box so that it surrounds only the portion of the image you want to keep.

8. **Tap the Done button to crop and save the image.**

9. **Save your changes: Tap the Done or Save button.**

   With some Contact apps, tap the Back navigation icon to exit Editing mode.

Table 7-1	Picture Confirmation Icons	
Icon	*Name*	*Function*
↰	Retry	Take another photo
✓	Done	Accept the photo and crop
✗	Cancel	Abandon your efforts

Save the cropped image   Portion kept

Resize rectangle   Portion discarded

Drag cropping rectangle

Figure 7-3: Cropping a contact's image.

The contact's image appears onscreen when the person calls, as well as when referenced in other apps, text messaging, Gmail, and so on.

To remove an image from a contact, follow Steps 1 and 2 in this section, but in Step 3 choose the action Remove Photo.

When you choose the action Select New Photo (Step 3), you can choose a new image for the contact, such as one stored on the phone, accessed through your Google Drive, or obtained from any number of photo storage or management apps.

- When the contact has no image, a generic image or icon appears.

- The contact image you see may be chosen by the contact themselves. Your Gmail contacts may show their own Google account image. Contacts associated with social networking sites may also use the person's own image instead of one you assign.

- Some images stored on the phone may not work for contact icons. For example, images synchronized with your online photo albums may be unavailable.

Making a favorite

A *favorite* contact is someone you stay in touch with most often. The person doesn't have to be someone you like — just someone you (perhaps unfortunately) phone often, such as your parole officer.

To create a favorite, display a contact's information and tap the Star icon. When the star is highlighted, the contact is flagged as one of your favorites.

Favorite contacts appear on their own tab in the People app or Contacts app, as well as in the Phone app. To quickly access a favorite, choose the person from the tab.

- To remove a favorite, tap the contact's star again. Removing a favorite doesn't delete the contact.

- A contact has no idea whether he's one of your favorites, so don't believe that you're hurting his feelings by not making him a favorite.

Joining identical contacts

Your phone can pull in contacts from sources such as Facebook, Gmail, Skype, and more. Because your friends also use multiple online services, you may discover several contact entries for the same person. When you notice such duplicates, you can join them. Here's how:

1. **Wildly scroll the address book until you locate a duplicate.**

   Well, maybe not *wildly* scroll, but locate a duplicated entry. Because the address book is sorted, the duplicates appear close together.

2. **Tap the Edit icon to edit one of the duplicate contacts.**

   Editing may not be required for some versions of the address book app.

   On some Samsung phones, tap the More icon and choose the Link Contacts action. Skip to Step 4.

3. **Tap the Action Overflow and choose Join.**

   The action might be titled Link Contact or something similar.

   After choosing the action, you see a list of contacts that the phone guesses could be identical. It also shows the entire list of contacts, in case the guess is incorrect. Your job is to find the duplicate contact.

4. **Choose the duplicate contact from the list.**

   The contacts are merged, appearing as a single entry in the address book.

Joined contacts aren't flagged as such in the address book, but you can easily identify them: A joined contact often lists two sources, such as Google and Facebook, in the Accounts area. Or you may see a Link icon (like a link in a chain).

Separating contacts

The topic of separating contacts has little to do with parenting, although separating bickering children is the first step to avoiding a fight. Contacts in the address book might not be bickering, but occasionally the phone may automatically join two contacts that aren't really the same person. When that happens, you can split them by following these steps:

1. **Display the improperly joined contact.**

   As an example, I'm Facebook friends with other humans named Dan Gookin. My phone accidentally joined my address book entry with another Dan Gookin.

2. **Tap the Action Overflow and choose Separate.**

   The command might not be available, in which case you need to edit the contact, as described earlier in this chapter. If so, tap the Action Overflow icon or MORE button to choose the Separate command.

   The Separate command might also be called Separate Contacts or Unlink Contacts.

3. **Tap the OK button to confirm that you're splitting the contacts.**

You don't need to actively look for improperly joined contacts — you'll just stumble across them. When you do, feel free to separate them, especially if you detect any bickering.

Removing a contact

Every so often, consider reviewing your phone's contacts. Purge the folks whom you no longer recognize or you've forgotten. It's simple:

1. **Edit the forlorn contact.**

2. **Tap the Delete icon.**

   If you don't see the Delete icon (shown in the margin), tap the Action Overflow to look for a Delete action.

   If you can't find any other logical way to remove the content, it was brought in from another source, such as Facebook. You need to use that app to disassociate the contact.

3. **Tap the OK button or Delete button to confirm.**

   Poof! They're gone.

Because the phone's address book is synchronized with your Google account, the contact is also removed there.

On the plus side, keep in mind that removing a contact doesn't kill the person in real life.

Text Messaging Mania

In This Chapter

▶ Creating a text message

▶ Getting a text message

▶ Forwarding a text message

▶ Texting pictures, videos, and media items

▶ Managing your text messages

▶ Specifying a text message ringtone

Texting is the popular name for using your Android phone to send short, typed messages to another phone — specifically, another cell phone. The process echoes earlier technology, such as the telegraph and teletype. Interestingly enough, the acronym LOL for *laugh out loud*, was first used in the 1880s:

```
Jeb: Stagecoach robbed again, but the boys
forgot the strongbox at the office. LOL.
```

Despite its seemingly anachronistic nature, texting remains a popular form of communications. Indeed, some young people text more than they use the phone to place a call. It's a great, convenient, and popular way to quickly communicate.

How 'bout them Patriots?
Aug 11, 14:35

Awesome
Aug 11, 14:36

Would you like some nuts?
Nov 2, 11:31

Are they in town again?
Nov 2, 11:36

Msg 4U

Text messaging allows you to send short quips of text from one cell phone to another. As long as the other phone is on and receiving a signal, the message is received instantly — not denying it! That makes texting worthy as a form of instant communication.

✔ Don't text while you're driving.

✔ Don't text in a movie theater.

✔ Don't text in any situation where it's distracting.

✔ Most cell phone plans include unlimited texting; however, some older plans may charge you per text. Check with your cellular provider to be sure.

✔ If you're over 25, you might want to know that the translation of this section's title is "Message for You."

✔ The nerdy term for text messaging is *SMS*, which stands for Short Message Service.

Selecting a texting app

Traditionally, Android phones come with one texting app. It's usually called Texting, although Messages, Messaging, and Text Messaging are also popular. The Google Hangouts app can also handle text messages, and it might be your phone's only texting app. Google's Messenger app is its preferred app for texting.

To confirm which texting app the phone uses, or to choose one or the other, follow these steps:

1. **Open the Settings app.**

   Choose the Settings app from the Apps drawer.

2. **Choose More.**

   This item is found at the bottom of the Wireless & Networks area on the Settings app's main screen.

3. **Choose Default SMS App.**

   This item might be titled Default Messaging App.

4. **Select a text messaging app.**

   If the only option available is Hangouts or Messenger, that's your SMS app.

You cannot disable Hangouts from being your phone's texting app when no other texting app is available.

Composing a new text message

The easiest way to start a new text message is to open the phone's address book app, choose a contact, and tap the Text Messaging icon next to the phone number. A new text message window instantly appears.

Whether to send a text message or an email

Sending a text message is similar to sending an email message. Both involve the instant electronic delivery of a message to someone else. And both methods of communication have their pros and cons.

The primary limitation of a text message is that it can be sent only to another cell phone. If you use Hangouts as your text messaging app, you can also send a text message to someone with a computer or tablet, but that person also has to use Hangouts.

Email is available to anyone who has an email address, whether or not the person uses a mobile communications device.

Text messages are pithy: short and to the point. They're informal, more like quick chats.

Indeed, the speed of reply is often what makes text messaging useful. As with email, however, sending a text message doesn't guarantee a reply.

An email message can be longer than a text message. You can receive email on just about any Internet-connected device. Email message attachments (pictures, documents) are handled better and more consistently than text message (MMS) media.

Finally, email is considered a wee bit more formal than a text message. But when you truly desire formal communications, make a phone call or send a letter.

To send a text message to a phone number not currently in the phone's address book, follow these steps:

1. **Open the phone's texting app.**

2. **At the app's main screen, tap the New Conversation button.**

   Frequently, the button is the Add (plus sign) icon. It might also be a Compose icon, as shown in the margin.

   In the Hangouts app, tap the Conversations tab. See Chapter 11 for specifics on the Hangouts app.

3. **Type the phone number.**

   As you type, matching contacts appear.

4. **In the Hangouts app, choose the SMS option from the list of matching phone contacts, as shown in Figure 8-1.**

   You must choose the SMS option; otherwise, you're starting a new Hangouts conversation. This is perhaps the only drawback to using Hangouts as the phone's messaging app.

5. **Type the message.**

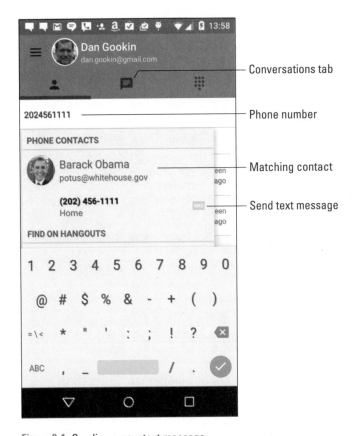

Conversations tab

Phone number

Matching contact

Send text message

Figure 8-1: Sending a new text message.

6. **Tap the Send icon to send the message.**

The Send icon may look similar to the one shown in the margin, or it may be only the word *SEND*.

- You can send text messages only to cell phones. Aunt Ida cannot receive text messages on her landline that she's had since the 1960s.

- You can send a single text message to multiple recipients: Just type additional phone numbers or contact names in the To field when you're composing a new message.

- When you receive a group message (one that has several recipients), you can choose whether to reply to everyone. Look for the Reply All button when composing your response.

- Do not text and drive. Do not text and drive. Do not text and drive.

Continuing a text message conversation

The text messaging app keeps track of old conversations. To continue a conversation at any time: Open the texting app, peruse the list of existing conversations, and tap one to review what has been said or to pick up the conversation.

Switching between Hangouts conversations and text messaging

Due to the Hangouts app's dual nature, you may have to switch between Hangouts conversations and pure text messaging or SMS mode. To do so, follow these steps:

1. **In the Hangouts app, tap the Side Menu icon to view the navigation drawer.**

2. **Tap your Gmail account at the top of the navigation drawer.**

3. **Choose SMS.**

   The Hangouts app switches to Text Messaging mode.

To switch back, tap the Side Menu icon and choose SMS again, and then tap your Gmail account.

Receiving a text message

New text messages are heralded by a notification atop the screen. The entire message may briefly appear on the status bar, but eventually all that remains is a New Text Message notification, similar to the one in the margin.

To view the message, pull down the notifications drawer. Tap the messaging notification, and that conversation window immediately opens.

Also see Chapter 11 for information on the Hangouts app's notification. When Hangouts is used as the text messaging app, its notification icon appears instead of a specific text-messaging icon.

Forwarding a text message

It's possible to forward a text message, but it's not the same as forwarding email. In fact, when it comes to forwarding information, email has text messaging beat by well over 160 characters.

Common text message abbreviations

Texting isn't about proper English. Indeed, many of the abbreviations and shortcuts used in texting, such as LOL and BRB, are slowly becoming part of the English language.

The weird news is that these acronyms weren't invented by teenagers. Sure, the kids use them, but the acronyms have their roots in the Internet chat rooms of yesteryear. Regardless of its source, you might find using a shortcut handy for typing messages quickly. Or maybe you can use this reference for deciphering an acronym's meaning. You can type acronyms in either uppercase or lowercase letters.

2	To, also
411	Information
BRB	Be right back
BTW	By the way
CYA	See you
FWIW	For what it's worth
FYI	For your information
GB	Goodbye
GJ	Good job
GR8	Great
GTG	Got to go
HOAS	Hold on a second
IC	I see
IDK	I don't know
IMO	In my opinion
JK	Just kidding
K	Okay
L8R	Later
LMAO	Laughing my ass off
LMK	Let me know
LOL	Laugh out loud
NC	No comment
NP	No problem
NRN	No reply needed (necessary)
OMG	Oh my goodness!
PIR	People in room (watching)
POS	Person over shoulder (watching)
QT	Cutie
ROFL	Rolling on the floor, laughing
SOS	Someone over shoulder (watching)
TC	Take care
THX	Thanks
TIA	Thanks in advance
TMI	Too much information
TTFN	Ta-ta for now (goodbye)
TTYL	Talk to you later
TY	Thank you
U2	You too
UR	You're, you are
VM	Voicemail
W8	Wait
XOXO	Hugs and kisses
Y	Why?
YW	You're welcome
ZZZ	Sleeping

When forwarding works as an option, your phone lets you forward only the information in a text messaging cartoon bubble, not the entire conversation. Here's how it works:

1. **If necessary, open a conversation in the phone's texting app.**

2. **Long-press the text entry (the cartoon bubble) you want to forward.**

3. **Choose Forward or Forward Message.**

   From this point on, forwarding the message works like sending a new message from scratch.

4. **Type the recipient's name (if the person is a contact), or type a phone number.**

   The text you're forwarding appears, already written, in the text field.

5. **Tap the Send icon to forward the message.**

Multimedia Messages

Even though the term *texting* sticks around, a text message can contain media — usually a photo — although short videos and audio can also be shared with a text message. Such a message ceases to be a mere *text* message and becomes a *multimedia message.*

- ✒ Multimedia messages are handled by the same app you use for text messaging.

- ✒ Not every mobile phone can receive multimedia messages. Rather than receive the media item, the recipient may be directed to a web page where the item can be viewed on the Internet.

- ✒ The official name for a multimedia text message is Multimedia Messaging Service, which is abbreviated MMS.

Creating a multimedia text message

As with other things on your Android phone, you need to think of sharing when it comes to attaching media to a text message. Obey these steps:

1. **Open the app that contains or shows the item you want to share.**

   For example, open the Photos app to view a picture or view a page in the web browser app.

2. **View the item and tap the Share icon.**

3. **Choose the text messaging app from the list of apps.**

4. **When Hangouts is your phone's text messaging app, choose the SMS option or account.**

5. **Continue sending the text message as described earlier in this chapter.**

 It's also possible to attach media to a message from within the text messaging app. To do so, begin composing the message, but look for the Attachment icon, similar to the one shown in the margin. Tap that icon, and then choose the media to attach.

In just a few, short, cellular moments, the receiving party will enjoy your multimedia text message.

Receiving a multimedia message

A multimedia attachment comes into your phone just like any other text message does. You may see a thumbnail preview of whichever media was sent, such as an image, a still from a video, or the Play icon to listen to audio. To preview the attachment, tap it.

 To do more with the multimedia attachment, long-press it and then select an action from the list. For example, to save an image attachment, long-press the image thumbnail and choose Save Picture.

Text Message Management

You don't have to manage your messages. I certainly don't. But the potential exists: If you ever want to destroy evidence of a conversation, or even do something as mild as change the text messaging ringtone, it's possible.

Removing messages

Although I'm a stickler for deleting email after I read it, I don't bother deleting my text message threads. That's probably because I have no pending divorce litigation. Well, even then, I have nothing to hide in my text messaging conversations. If I did, I would follow these steps to delete a conversation:

1. **Open the conversation you want to remove.**

   Choose the conversation from the main screen in your phone's text messaging app.

2. **Tap the Action Overflow and choose Delete.**

3. **Tap the Delete button to confirm.**

    The conversation is gone.

If these steps don't work, an alternative is to open the main screen in the text messaging app and long-press the conversation you want to zap. Tap the Delete button, and then tap the Delete or OK button to confirm.

Setting the text message ringtone

The sound you hear when a new text message floats in is the text message ringtone. It might be the same sound you hear for all notifications, though on some Android phones it can be changed to something unique.

In the Hangouts app, follow these steps to set a new text-message ringtone (which is also the Hangouts ringtone):

1. **Tap the Side Menu icon to display the navigation drawer.**

2. **Choose Settings.**

3. **Tap your Gmail account.**

4. **Choose Hangouts Messages & Invites.**

5. **Choose Sound.**

6. **Select a sound from the list and tap OK.**

If you're using a specific text messaging app (not Hangouts), open that app and tap the Action Overflow icon or MORE button. Choose Settings, and then look for the Notifications or Ringtone item. Eventually, you see an item from which you can select a new ringtone.

9

Email This and That

*T*he first official telegraph message was, "What hath God wrought?" The first telephone call was supposedly, "Mr. Watson. Come here. I want you." The first email message reportedly said, "Wànt tø büy somé chéàp Ciàlis?"

You'll be delighted to know that your phone can send and receive email. You can ask questions about the Almighty, receive requests for assistance, and trade in pharmaceuticals.

Air Force One

I need to be in Albuquerque Tues can you give me a lift?

Email on Your Android Phone

For most Android phones, two apps handle email: Gmail for Google email, and the Email app for everything else — including web-based email such as Yahoo! Mail or Windows Live mail, and even corporate mail.

Recently, however, Google has expanded the Gmail app to cover all types of email. This may be the case on your phone, or you might still find separate Gmail and Email apps. The good news is that both apps work in a similar manner, so your sanity should remain more or less intact.

✔ The Gmail app is updated frequently. To review any changes since this book went to press, visit my website at:

    `www.wambooli.com/help/android`

✔ Although you can use the phone's web browser app to visit the Gmail website, you should instead use the Gmail app to pick up your Gmail. Likewise, you can use a web page to read most other email, but using the Email (or newer Gmail) app is preferred.

✔ If you forget your Gmail password, visit this web address:

    `www.google.com/accounts/ForgotPasswd`

Setting up the first Email account

If your phone still has the Email app, getting that first account set up works differently than when you're adding additional accounts. These steps apply only to the Email app. If you're using the Gmail app to add more email accounts, refer to the next section.

The key to the painless addition of your first email account is to know the email address and password. If you're good to go, follow these steps:

1. **Start the Email app.**

   Look for it in the Apps drawer. If it's not there, use the Gmail app instead. See the next section.

   The first screen you see is Set Up Email. If you've already run the Email app, you're taken to the Email inbox and you can skip these steps. See the next section for information on adding accounts.

2. **Type the email address you use for the account.**

   For example, if you have a Comcast email account, use the onscreen keyboard to type your `me@comcast.net` email address in the Email Address box.

   If you see the .com key on the onscreen keyboard, use it to more efficiently type your email address.

3. **Type the account's password.**

4. **Tap the Next button on the screen or the Done button on the onscreen keyboard.**

   If you're lucky, everything is connected and you can move on to Step 5. Otherwise, you have to specify the details as provided by your ISP. See the later section "Adding an account manually."

5. **Set account options.**

   You might want to reset the Inbox Checking Frequency to something other than 15 minutes. I recommend keeping the other items selected until you become familiar with how your phone handles email.

6. **Tap the Next button.**

7. **Give the account a name and confirm your own name.**

   The account is given the name of the mail server or your email address. That choice may not ring a bell when you receive your email. I name my ISP's email account Main because it's my main account.

   The Your Name field shows your name as it's applied to outgoing messages. So if your name is really Annabelle Leigh Meriwether and not amer82, you can make that change now.

8. **Tap the Done button.**

   You're done.

The next thing you see is your email account inbox. The phone proceeds to synchronize any pending email you have in your account, updating the screen as you watch. See the later section "You've Got Mail" for what to do next.

 If you use Yahoo! Mail, I recommend getting the Yahoo! Mail app, which handles your Yahoo mail far better than the Email app (or the Gmail app). The Yahoo! Mail app gives you access to other Yahoo features that you may use and enjoy. Obtain the Yahoo! Mail app from the Google Play Store. See Chapter 16.

Adding more email accounts

The Email app, as well as the newest version of Gmail, can be configured to pick up email from multiple sources. If you have a Windows Live account or maybe an evil corporate account in addition to your ISP's account, you can add them.

Both the Email and Gmail apps offer different ways to add a new email account. A more generic approach is to use the Settings app. Obey these directions:

1. **Open the Settings app.**

   It's found in the Apps drawer; tap the Apps icon on the Home screen to view the Apps drawer.

2. **Choose Accounts.**

   On some Samsung phones, tap the General tab to locate the Accounts item.

3. **Tap Add Account.**

4. **Choose the email account type.**

   The three options for adding email accounts are

   Exchange or *Microsoft Exchange ActiveSync:* For a corporate email account hosted by an Exchange Server (Outlook mail)

   Personal (IMAP): For web-based email accounts, such as Microsoft Live

   Personal (POP3): For traditional, ISP-email accounts, such as Comcast

   See the later section "Adding a corporate email account" for information on the Exchange option. The other two options work pretty much the same. In fact, some phones may have a single Email option instead of the IMAP and POP variations.

5. **Type your email address and tap the Next button.**

6. **Type the email account password and tap the Next button.**

7. **Continue working through the email setup as described in the preceding section.**

   Start with Step 5, where you review the account settings.

The big difference between creating the first email account and adding more is that you're asked whether the account you just added is the primary or default account. See the later section "Setting the primary email account," for details.

The new email account is synchronized immediately after it's added, and you see the inbox. See the later section "Checking the inbox."

Adding an account manually

If your email account isn't recognized, or some other booboo happens, you have to manually add the account. The steps in the preceding sections remain the same, but you need to provide more specific and technical details. This information includes tidbits such as the server name, port address, domain name, and other bothersome details.

My advice is to contact your ISP or email provider: Look on their website for specific directions on adding their email service to an Android phone. Contact them directly if you cannot locate specific information.

The good news is that manual setup is very rare these days. Most ISPs and webmail accounts are added painlessly, as described earlier in this chapter.

Adding a corporate email account

The easiest way to set up your evil corporation's email on your phone is to have the IT people do it for you. Or you may be fortunate, and on the organization's intranet you'll find directions. I present this tip because configuring corporate email, also known as Exchange Server email, can be a difficult and terrifying ordeal.

It's possible to add the account on your own, but you still need detailed information. Specifically, you need to know the domain name, which may not be the same as the outfit's website domain. Other details may be required as well, the details of which I shudder to think about.

Above all, you need to apply a secure screen lock to your phone to access Outlook email. This means you need to add a PIN or password to the device, which is covered in Chapter 21. You cannot access the Exchange Server without that added level of security.

And there's more!

You also need to grant Remote Security Administration privileges. This means your organization's IT gurus will have the power to remotely wipe all information from your phone. You must activate that feature, which is part of the setup process.

The bonus is that when you're done, you have full access to the Exchange Server info. That includes your email messages as well as the corporate address book and calendar. Refer to Chapter 7 for information on the address book app; the Calendar app is covered in Chapter 15.

You've Got Mail

New email arrives into your phone automatically, picked up according to the Gmail and Email apps' synchronization schedules. If Gmail is the phone's only email app, you use it to read all your email. Otherwise, use the Email app to read non-Gmail email.

Getting a new message

You're alerted to the arrival of a new email message in your phone by a notification icon. The icon differs depending on the email's source.

For a new Gmail message, the New Gmail notification appears, similar to the one shown in the margin.

For a new email message, you see the New Email notification.

Conjure the notification drawer to review the email notifications. Tap a notification to be whisked to an inbox for instant reading, as described in the next section.

Checking the inbox

To peruse your Gmail, start the Gmail app. A typical Gmail inbox is shown in Figure 9-1.

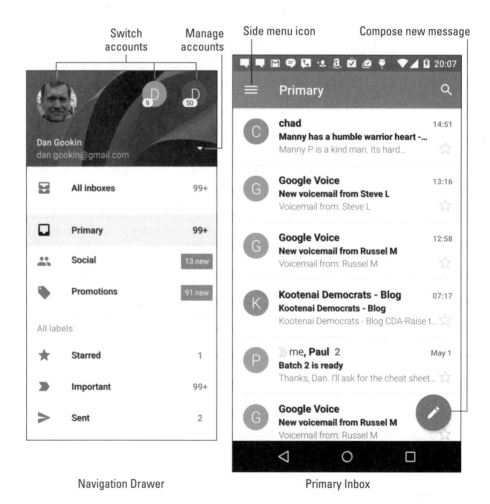

Navigation Drawer   Primary Inbox

Figure 9-1: The Gmail inbox.

Choose non-Gmail accounts by tapping the account icon from the navigation drawer, shown in Figure 9-1. You can view only one account's inbox at a time.

If your phone still uses the Email app, open it to view its inbox. You either see a single account's inbox or you can choose to view the universal inbox, shown as Combined inbox in Figure 9-2.

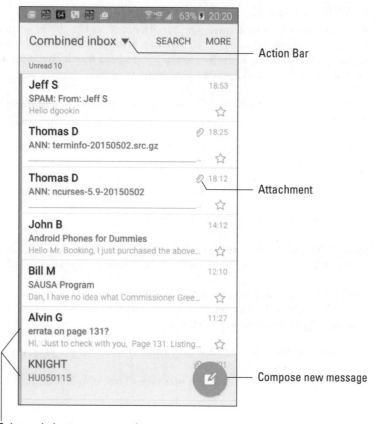

Figure 9-2: Messages in the Email app.

✏ Gmail doesn't show up in the Email app, not even in the Combined inbox. Use the Gmail app to read your Google mail.

✏ The Gmail app lacks a combined inbox. To view specific inboxes, tap an account bubble on the navigation drawer, as illustrated earlier in Figure 9-1.

✔ Multiple email accounts that are gathered in the Email app are color coded. When you view the combined inbox, you see the color codes to the left of each message, as shown in Figure 9-2.

✔ To view an individual account's inbox in the Email app, choose the account from the action bar or from the navigation drawer.

Reading email

To view a specific email message, tap its entry on the inbox screen, as shown earlier, in Figures 9-1 and 9-2. Or you can choose a new email notification. Reading and working with the message operates much the same whether you're using the Gmail app or Email app:

✔ Swipe the message up or down by using your finger to read it.

✔ Browse between messages by swiping the screen left or right.

✔ Tap the left-pointing arrow or chevron at the top left corner of the screen to return to the inbox.

To work with the message, use the icons that appear above or below the message text. These icons, which may not look exactly like those shown in the margin, cover common email actions:

 Reply: Tap this icon to reply to a message. A new message window appears with the To and Subject fields reflecting the original sender(s) and subject.

 Reply All: Tap this icon to respond to everyone who received the original message, including folks on the Cc line. Use this option only when everyone else must get a copy of your reply.

 Forward: Tap this icon to send a copy of the message to someone else.

 Delete: Tap this icon to delete a message. This icon is found atop the message.

● If you don't see all of these icons — specifically, the Reply All and Forward
● icons — change the phone's orientation to horizontal. You might also find all
● the actions by tapping the Action Overflow.

 If you've properly configured the Email program, you don't need to delete messages you've read. See the section "Configuring the server delete option," later in this chapter.

Make Your Own Email

Although I use my phone often to check email, I don't often use it to compose messages. That's because most email messages don't demand an immediate reply. When they do, or when the mood hits me and I feel the desperation that only an immediate email message can quench, I compose a new email message on my Android phone using the methods presented in this section.

Writing a new electronic message

Creating a new email epistle works similarly with both the Gmail and Email apps. The key is to tap the Compose icon. Tap that icon and then fill in the blanks, adding recipient, subject, and message text.

Figure 9-3 shows the layout of the email composition screen, which should be familiar to you if you've ever written email on a computer. The screen is similar between the Gmail and Email apps, although not identical.

To send the message, tap the Send icon, illustrated in Figure 9-3 and shown in the margin.

- ✔ To choose an email account for sending, tap the chevron next to your name atop the Compose screen, shown in Figure 9-3. Otherwise, the primary or default email account is used. See the later section "Setting the primary email account."

- ✔ You need only type a few letters of the recipient's name. Matching contacts from the phone's address book appear in a list. Tap a contact to automatically fill in the To field.

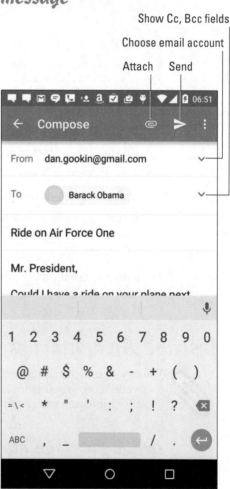

Show Cc, Bcc fields

Choose email account

Attach   Send

Figure 9-3: Writing a new email message.

- To view the CC and BCC fields, tap the chevron, shown in Figure 9-3. If that doesn't work, tap the Action Overflow and choose Add Cc/Bcc.

- To cancel a message, tap the Action Overflow and choose Discard. Tap the OK or Discard button to confirm.

- Save a message by choosing the Save Draft action from the Action Overflow. Drafts are saved in the Drafts folder. You can open them there for further editing or sending.

- Some phones feature a formatting toolbar in the Email app. Use it to change the way the text looks in your message.

Sending email to a contact

A quick and easy way to compose a new message is to find a contact in the phone's address book app. Heed these steps:

1. **Open the phone's address book app.**

   See Chapter 7 for details on using this app.

2. **Locate the contact to whom you want to send an electronic message.**

3. **Tap the contact's email address.**

4. **Choose Gmail to compose the message.**

   If the Email app is available, or another app such as Yahoo! Mail, choose it instead. Also see Chapter 19 for information on choosing a default app.

At this point, creating the message works as described in the preceding sections.

Message Attachments

The key to understanding email attachments on your phone is to look for the paperclip icon. When you see that icon, you can either deal with an attachment for incoming email or add an attachment to outgoing email.

Dealing with an attachment

Attachments are presented differently between the Gmail and Email apps. Either way, your goal is the same: to view the attachment or to save it. Sometimes you can do both!

Figure 9-4 shows how the Gmail and Email apps can present an email attachment. To deal with the attachment, tap it. In most cases, the attachment opens an appropriate app. For example, a PDF attachment might be opened by the Quickoffice app.

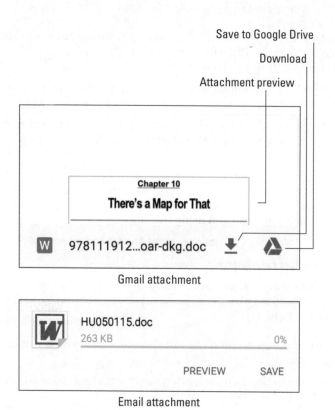

Figure 9-4: Attachment methods and madness.

Potential actions you can perform with an attachment include

Preview: Open the attachment for viewing.

Save to Google Drive: Send a copy of the attachment to your Google Drive (cloud) storage.

Save: Save the attachment to the phone's storage.

Download Again: Fetch the attachment from the mail server.

As with email attachments received on a computer, the only problem you may have is that the phone lacks the app required to deal with the attachment. When an app can't be found, you either have to suffer through not viewing the attachment or simply reply to the message and direct the person to resend the attachment in a common file format.

✔ Common file formats include PNG and JPEG for pictures, and HTML or RTF for documents. PDF and DOCX for Adobe Acrobat and Microsoft Word documents are also common.

✔ See Chapter 19 for information on selecting a default app should the phone prompt you to choose an app to view the attachment.

✔ Images in an email are technically attachments, although they're sometimes not displayed. Tap the message's Show Pictures button to see the images.

✔ Attachments are saved in the Downloads or Download folder on the phone's internal storage. You can view that folder by using the Downloads app, which is covered in Chapter 10. Also see Chapter 18 for information on phone storage.

Sharing an attachment

Although it's possible to attach something to an email message on your phone, the preferred way to create an email attachment is to use the Share icon. Follow these steps:

1. **Open the app that created the attachment, or in which you can view the attachment.**

   For example, to send a photo, open the Photos app. To send a YouTube video link, open the YouTube app.

2. **View the item you want to share.**

3. **Tap the Share icon.**

   A list of apps appears.

4. **Choose Email or Gmail.**

   This prompt may not appear if you've chosen a default email app.

5. **Compose the message.**

   The item is attached to the message, so at this point writing the email works just as described earlier in this chapter.

The computer way of attaching a file to a message is to first use the Gmail or Email apps and write the message. Tap the Action Overflow icon and choose the command Attach File, or tap the Attachment icon as shown in the margin. Use the screen that appears to choose an app, or browse through the files stored on the phone to pick one as an attachment.

   ✔ The variety of items you can attach depends on which apps are installed on the phone.

   ✔ The Gmail and Email apps sometimes accept different types of attachments. So if you cannot attach something by using the Gmail app, try using the Email app instead.

Email Configuration

You can have oodles of fun and waste oceans of time confirming and customizing the email experience on your Android phone. The more interesting things you can do are to modify or create an email signature, specify whether messages retrieved on the phone are deleted from the server, and assign a default email account.

Creating a signature

I highly recommend that you create a custom email signature for sending messages from your phone. Here's my signature:

```
DAN
This was sent from my Android phone.
Typos, no matter how hilarious, are unintentional.
```

To create a custom signature for your email accounts, obey these directions:

1. **Open the Settings app**

2. **Choose Accounts.**

   Some Samsung phones hide the Accounts item on the General tab.

3. **Choose an email account from the list.**

   For example, choose the Personal (IMAP) or Personal (POP3) items, or tap the Email item if you see only it.

4. **Choose Account Settings.**

   This item might be titled Settings.

5. **Choose a specific email account.**

6. **Choose Signature.**

   Any existing signature appears on the card, ready for you to edit or replace it.

7. **Type or dictate your signature, and then tap the OK button.**

   The signature you set is appended automatically to all outgoing email you send from that account.

You need to repeat these steps (actually, only Steps 5–7) for each of your email accounts.

If you want to use the same signature on other accounts, select and copy the text from Step 7. See Chapter 4 for details on text editing.

These steps may not work on your phone's Email app. If so, open that app and locate the Settings item. (Tap the Action Overflow or the MORE button.) Choose your email account from the list, and then choose Signature to create a custom signature.

Configuring the server delete option

Unlike reading your email on a computer, messages you fetch from traditional ISPs (such as Comcast or CenturyLink) on your phone aren't deleted from the email server. The advantage is that you can retrieve the same messages later by using a computer. The disadvantage is that you end up retrieving mail you've already read and possibly replied to.

You can control whether messages are removed after they're picked up. This setting applies only to POP3 email, not Gmail or IMAP (webmail) accounts. Follow these steps:

1. **Open the Settings app and choose Accounts.**

2. **Tap the Personal (POP3) item.**

   This item might simply be titled Email.

3. **Choose Account Settings.**

   The action might be titled Settings.

4. **Choose an email account.**

5. **Choose Incoming Settings.**

   Look for a More Settings button if you don't directly see the Incoming Settings item.

6. **Tap the Delete Email from Server item.**

7. **Choose the When I Delete from Inbox option.**

8. **Tap the Done button.**

When these steps don't work, use the Email app to change the setting: Tap the Action Overflow (or MORE button), and choose Settings. Continue with Step 4 from the list.

After you make or confirm this setting, messages you delete on the phone are also deleted from the mail server. That means the message won't be picked up again, not by the phone, another mobile device, nor any other computer that fetches email from that same account.

Setting the primary email account

When you have more than one email account, the main account — the *default* — is the one used by the Email app to send messages. To change that primary mail account, follow these steps:

1. **Start the Email app.**

2. **Choose Combined View or Combined Inbox.**

   This option is chosen from an action bar or the navigation drawer.

3. **Tap the Action Overflow or MORE button.**

4. **Choose Settings.**

5. **Tap the email account you want to mark as the primary or default account.**

6. **Select the Default Account item.**

   The messages you compose and send using the Email app are sent from the account you specified in Step 5.

When you can't find a Default Account item (Step 6), you can always choose an outgoing email account when you compose a new message as described earlier in this chapter.

10

Out On the Web

*W*hen Tim Berners-Lee developed the World Wide Web back in 1990, he had no idea that people would one day use it on their cell phones. Nope, the web was designed to be viewed on a computer, specifically one with a nice, roomy, high-resolution monitor and a full-size keyboard. Cell phones? They had teensy LED screens. Browsing the web on a cell phone would have been like viewing the *Mona Lisa* through a keyhole.

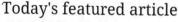

Well, okay: Viewing the web on your cell phone *is* kind of like viewing the *Mona Lisa* through a keyhole. The amazing thing is that you can do it in the first place. For some web pages, it's workable. For others, you have my handy advice in this chapter.

Today's featured article

Tiruchirappalli is the fourth largest city in the Indian sta...

m.wambooli.com   🌐 Wambooli Mobi

Welcome to Wambooli Mobile

~ 19. 2014 on ~

✔ If possible, activate the phone's Wi-Fi connection before you venture out on the web. Although you can use the mobile data connection, a Wi-Fi connection incurs no data usage charges.

✔ Many places you visit on the web can instead be accessed directly and more effectively by using specific apps. Facebook, Gmail, Twitter, YouTube, and other popular web destinations have apps that you may find are already installed on your phone or otherwise available for free from the Google Play Store.

> ✔ One thing you cannot do with your phone is view Flash animations, games, or videos. The phone's web browser disables the Flash software, also known as Shockwave. I know of no way to avoid this limitation.

The Phone's Web Browser App

All Android phones feature a web browsing app. The stock Android app is Google's own Chrome web browser. Your phone may use another web browser app, such as an app named Web, Browser, or Internet.

The good news is that all web browser apps work in a similar way and offer comparable features. The secret news is that most of these other web browsers are simply gussied-up versions of the Chrome web browser. And the sobering news is that I'm writing about only one web browser app in this chapter, which is the Chrome app.

> ✔ If your phone doesn't have the Chrome app, you can obtain it for free at the Google Play Store. See Chapter 16.
>
> ✔ An advantage of using Chrome is that your bookmarks, web history, and other features are shared between all devices you use on which Chrome is installed. So if you use Chrome as your computer's web browser, it's logical to use Chrome on your phone as well.
>
> ✔ The first time you fire up the web browser app on certain Samsung phones, you may see a registration page. Register your device to receive sundry Samsung bonus stuff — or not. Registration is optional.

Behold the Web

Rare is the person these days who has had no experience with the World Wide Web. More common is someone who has used the web on a computer but has yet to taste the Internet waters on a mobile device. If that's you, consider this section your quick mobile-web orientation.

Surfing the web on your phone

When you first open the web browser app, you're taken to the home page. The Chrome app lacks a home page, so instead you see the last page you were viewing. Or, if you're starting the app for the first time, you see Google's main page, illustrated in Figure 10-1.

Here are some handy phone web-browsing tips:

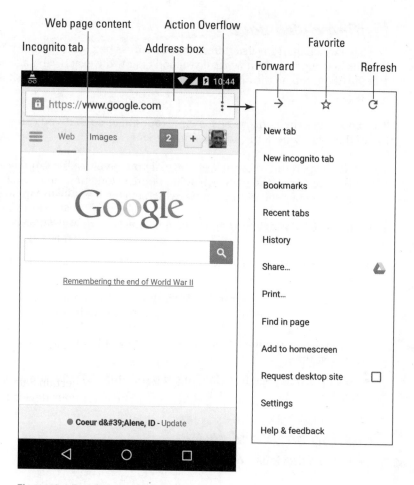

Figure 10-1: The Chrome app beholds the Google home page.

- ✔ *Pan* the web page by dragging your finger across the touchscreen. You can pan up, down, left, and right, which comes in handy when the web page is larger than the phone's screen.

- ✔ *Pinch* the screen to zoom out, or *spread* two fingers to zoom in.

- ✔ The page you see may be the mobile page, or a customized version of the web page designed for small-screen devices. To see the nonmobile version, tap the Action Overflow icon and choose Request Desktop Site (refer to the right side of Figure 10-1).

- ✔ You can reorient the phone horizontally to read a web page in Landscape mode.

Visiting a web page

To visit a web page, type its address in the address box (refer to Figure 10-1). You can also type a search word, if you don't know the exact web page address. Tap the Go button on the onscreen keyboard to search the web or visit a specific web page.

If you don't see the address box, swipe your finger so that you can see the top of the page, where the address box lurks.

Tap links to open the relevant web page. If you have trouble tapping a specific link, zoom in on the page. On some phones, long-pressing the screen may display a magnification window, which makes it easier to tap links.

- ✓ The onscreen keyboard may change some keys to make it easier to type a web page address. Look for a .www or .com (dot-com) key.

- ✓ Long-press the .com key to see other top-level domains, such as .org or .net.

- ✓ To reload a web page, tap the Refresh icon. If you don't see that icon on the screen, tap the Action Overflow to find the Refresh or Reload action. Refreshing updates a website that changes often. Using the Refresh action can also reload a web page that may not have completely loaded the first time.

- ✓ To stop a web page from loading, tap the Stop (X) icon that appears by the Address box.

Browsing back and forth

To return to a previous web page, tap the Back navigation icon.

To go forward after going back, tap the web browser app's Forward icon. This icon is found on the Action Overflow, as illustrated in Figure 10-1, although some web browser apps may place it next to the address box.

To review web pages you've visited, visit the web browser's history page. In Chrome, choose History from the Action Overflow. If your phone doesn't use Chrome, try looking on the Bookmarks page for your browsing history: Tap the Action Overflow and choose Bookmarks.

- ✓ See the later section "Clearing your web history" for information on purging items from the History list.

- ✓ If you find yourself frequently clearing the web page history, consider using an incognito tab. See the later section "Going incognito."

Working with favorite pages (bookmarks)

You may know them as bookmarks on a computer web browser, but on your phone those special pages you save for quick access later are called *favorites*. To make a web page one of your favorites, follow these steps:

1. **Navigate to the web page.**

   Favorite web pages are those you plan on visiting frequently or want to return to later.

2. **Tap the Action Overflow and choose the Star icon.**

   The Star (Favorite) icon is shown in the margin, as well as earlier in Figure 10-1. Upon success, you see the Add Bookmark card, similar to the one shown in Figure 10-2.

3. **Edit the bookmark's name, or type in a better name.**

   I prefer short, punchy names as opposed to the long, meandering text that often titles a web page.

4. **If you're an organized person, tap the Folder action bar to set a specific folder for the bookmark.**

   The action bar is shown on the left in Figure 10-2. The folders then appear, as shown on the right. Tap a folder to choose a place for the bookmark. If you're an organized person, you'll probably have a complex hierarchy of folders to choose from. If you're not organized, well then, who cares?

5. **Tap the Save button to create the bookmark.**

To view bookmarks in the Chrome app, tap the Action Overflow and choose Bookmarks. You see the Bookmarks screen. Tap a folder to browse bookmarks; tap a bookmark to visit that page.

To quickly visit a bookmarked web site, just start typing the site's name in the address box. Tap the bookmarked site from the matching list of results displayed below the Address box.

- ✔ The Desktop bookmarks folder lists bookmarks shared with the computer version of Chrome. The Mobile Bookmarks folder lists bookmarks you've created when using a mobile device. My advice is to save all your bookmarks in the Desktop bookmarks folder.

- ✔ Remove a bookmark by long-pressing its entry on the Bookmarks screen. Choose the Delete Bookmark action.

- ✔ I confess that bookmarks are easier to manage when using a computer.

- ✔ A great way to find which sites to bookmark is to view the web page history. See the preceding section.

Set bookmark folder

Edit name

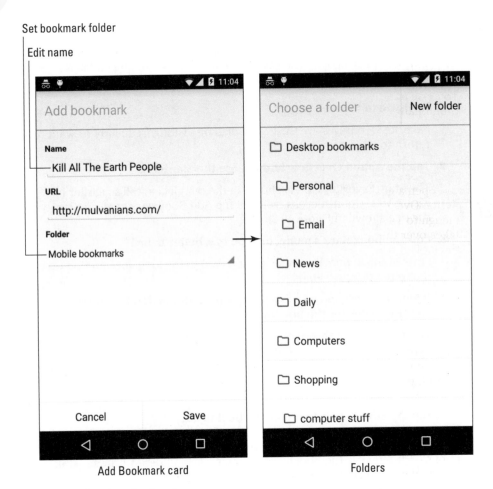

Add Bookmark card   Folders

Figure 10-2: Creating a bookmark.

Managing multiple web pages in tabs

Like many popular web browsers, the Chrome app uses a tabbed interface to help you access more than one web page at a time. You can do various interesting things with tabs:

- ✔ **To open a link in another tab,** long-press the link and choose Open in New Tab.

- ✔ **To open a bookmark in a new tab,** long-press the bookmark and choose Open in New Tab.

The problem is that the tabs aren't visible or easily accessible on your phone's screen. To see them all, tap the Action Overflow and choose New Tab. You see a list of open tabs, as shown in Figure 10-3. Tap a tab thumbnail to switch.

To close a tab, long-press its thumbnail and choose Remove.

Going incognito

Shhh! For secure browsing, you can open an *incognito* tab: Tap the Action Overflow and choose New Incognito Tab. The Incognito tab takes over the screen, changing the look of the Chrome app and offering a description page.

When you go incognito, the web browser doesn't track your history, leave cookies, or provide other evidence of which web pages you've visited. It doesn't prevent viruses or sophisticated snooping software.

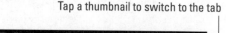

Tap a thumbnail to switch to the tab

✔ You cannot switch between incognito tabs, although you can open new ones by repeating the procedure mentioned in this section.

✔ Close one or all incognito tabs by choosing the Incognito notification, similar to the one shown in the margin.

Figure 10-3: Switching tabs.

Searching the web

The handiest way to find things on the web is to use the Google Search widget, found floating on the Home screen. You can also use Google Now, which is covered in Chapter 15. And, of course, you can visit Google's main search page by using the web browser app.

Or, what-the-hey: Be different and use Bing: www.bing.com.

Finding text on a web page

To locate text on a web page, tap the Action Overflow icon and choose Find in Page. Type the search text into the Find on Page box. As you type, found text is highlighted on the screen. Use the up or down chevrons to the right of the search box to page through the document.

Tap the Back icon to dismiss the Find in Page box after you've finished searching.

Sharing a web page

There it is! That web page you just *have* to talk about to everyone you know. The gauche way to share the page is to copy and paste it. Because you're reading this book, though, you know the better way to share a web page. Heed these steps:

1. **Go to the web page you desire to share.**

   Actually, you're sharing a *link* to the page, but don't let my obsession with specificity deter you.

2. **Tap the Action Overflow icon and choose Share.**

   The action might also be called Share Page or Share Via. Either way, a long list of apps is displayed. The variety and number of apps depends on what's installed on your phone.

3. **Choose an app.**

   For example, choose Gmail to send the link by email, or choose Facebook to share the link with your friends.

4. **Do whatever happens next.**

   Whatever happens next depends on how you're sharing the link: Compose the email, write a post in Facebook, or whatever. Refer to various chapters in this book for specific directions.

 You cannot share a page you're viewing by way of an Incognito tab.

Also see Chapter 18 for information on printing web pages.

The Joy of Downloading

Downloading is an easy concept to grasp, as long as you accept how most people misuse the term. Officially, a *download* is a transfer of information over a network from another source to your phone. For the web browser app, the network is the Internet, and the other source is a web page.

✔ The Download Complete notification appears after your phone has downloaded something. You can choose that notification to view the downloaded item.

✔ New apps are installed on your phone by using the Play Store app, covered in Chapter 16. That's a type of downloading, but it's not the same as the downloading described in this section.

✔ Most people use the term *download* when they really mean *transfer* or *copy.* Those people must be shunned.

✔ The opposite of downloading is *uploading.* That's the process of sending information from your gizmo to another location on a network.

Grabbing an image from a web page

The simplest thing to download is an image from a web page. Obey these steps:

1. Long-press the image.

You see an action card with a list of things you can do.

2. Choose Save Image.

The image is copied and stored on your phone.

See the later section "Reviewing your downloads," for details on how to access the image. It also appears in the Photos or Gallery app, which is covered in Chapter 13.

Downloading a file

The web is full of links that don't open in a web browser window. For example, some links automatically download, such as links to PDF files or Microsoft Word documents or other types of files that a web browser is too a-feared to display.

To save other types of links that aren't automatically downloaded, long-press the link and choose the Save Link action. If this action doesn't appear, your phone is unable to save the link, either because the file is of an unrecognized type or it's a security issue.

Reviewing your downloads

To review items downloaded from the Internet, choose the Download Complete notification, similar to the one shown in the margin.

When the Download notification goes away, you can access all downloaded material by opening the Downloads app. You'll find it eagerly awaiting your attention in the Apps drawer.

The Downloads app lists downloaded items organized by date. To view a download, tap an item in the list. The phone then starts the appropriate app to view the item.

- ✔ Some downloads you cannot view. That's because not every file is accessible on a mobile device.

- ✔ Technically, the image is stored on the phone's internal storage. It can be found in the Download or Downloads folder. You can read about file storage in Chapter 18.

Web Controls and Settings

More options and settings and controls exist for the web browser app than just about any other Android app I've used. Rather than bore you with every dang-doodle detail, I thought I'd present just a few of the options worthy of your attention.

Clearing your web history

When you don't want the entire Internet to know what you're looking at on the web, open an Incognito tab, as described earlier in the section "Going incognito." When you forget to do that, follow these steps to clear one or more web pages from the browser history:

1. **Tap the Action Overflow and choose History.**

2. **Tap the X icon next to the web page entry you want to remove.**

   It's gone.

You don't need to clean up your web browsing history when you use an incognito tab.

Changing the way the web looks

You can do a few things to improve the way the web looks on your phone. The key is to find the actions that adjust the text size or zoom level. Heed these steps:

1. **Tap the Action Overflow and choose Settings.**

2. **Choose Accessibility.**

The action might be also titled Screen and Text. I've also seen the action Default Zoom.

3. **Use the Text Scaling slider to adjust the text size.**

   The preview text below the slider helps you gauge which size works best.

  ✔ You can spread your fingers to zoom in on any web page.

  ✔ You can also turn the phone sideways to view a web page larger in the horizontal orientation.

Setting privacy and security options

No matter which web browser app you use, the security settings are most likely set to a comfortable level of protection. One item you might consider checking is how information is retained and automatically recalled for the web pages you visit. You may want to disable some of those features. Obey these steps:

1. **Tap the Action Overflow and choose Settings.**

2. **Choose Privacy.**

3. **Tap the Clear Browsing Data button.**

   This step might not be necessary for all web browser apps, which directly show you the list of check mark items.

4. **Place check marks by those items you want removed from the phone's storage.**

5. **Tap the Clear button.**

Some web browser apps might present options for remembering form data or passwords. If so, remove the check marks by those items. Also, delete any personal data, if that's an option.

With regard to general online security, my advice is always to be smart on the web and think before doing anything questionable. Use common sense. One of the most effective ways that the Bad Guys win is by using *human engineering* to try to trick you into doing something you normally wouldn't do, such as click a link to see a cute animation or a racy picture of a celebrity or politician. As long as you use your noggin, you should be safe.

Also see Chapter 21 for information on applying a secure screen lock, which I highly recommend.

Digitally Social

*S*ocial networking is that 21st century phenomenon that proves many odd beliefs about people. For example, it's possible to have hundreds of friends and never leave your house. You can jealously guard your privacy against the wicked intrusions of the government, all while letting everyone on the Internet know that you've just "checked in" to Starbucks with your friend Jane to have an iced mocha Frappuccino. And you can share your most intimate moments with humanity, many of whom will "like" the fact that you've just broken up or that your cat was run over by the garbage collection service.

🖊 Status   📷 Photo   📍 Cl

Linda C
Yesterday at 17:44

Think the Hawks will notice number 7 can ru

4 likes 8 comments

Like   💬 Comment   ➔ S

Share Your Life on Facebook

Of all the social networking opportunities, Facebook is the king. It's the online place to go to catch up with friends, send messages, express your thoughts, share pictures and videos, play games, and waste more time than you ever thought you had.

> ✔ The best way to access Facebook is to use the Facebook app. This app is preinstalled on some phones. If not, you can obtain the Facebook app for free at the Google Play Store. See Chapter 16.

> ✔ You can use the Facebook app to sign up for a Facebook account, or you can sign in using your existing account.

✔ After signing in to Facebook the first time, you have to perform configuration. I recommend choosing the option to synchronize Facebook to
your phone's address book.

✔ The Facebook app is frequently updated. Review any new information by
visiting my website:

        www.wambooli.com/help/android

Using Facebook on your phone

The Facebook app presents information on several tabs, as shown in
Figure 1-1. The tab you'll probably use the most is the News Feed. Options
for interacting with Facebook appear near the top of the News Feed, as illustrated in the figure.

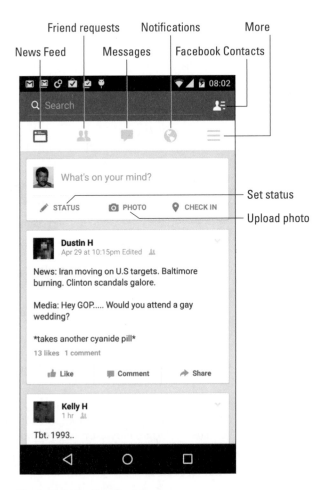

Figure 11-1: The Facebook app.

The Facebook app continues to run until you either sign out or turn off the phone. To leave the app, tap the Home navigation icon. Or you can tap the Recent navigation icon to switch to another, running app.

To sign out of the Facebook app, tap the More icon (refer to Figure 11-1) and choose the Log Out action. (Swipe down the screen to locate that action.) Tap the Log Out button to confirm.

> ✔ Refer to Chapter 19 for information on placing a Facebook launcher or widget on the Home screen.

> ✔ Swipe the news feed downward to instantly refresh its contents.

> ✔ Use the Like, Comment, or Share buttons below a News Feed item to comment, like, or share something, respectively. You can see any existing comments only when you choose the Comment item.

> ✔ The Facebook app generates notifications for new news items, mentions, chat, and so on. The Facebook notification icon looks similar to the one shown in the margin.

Setting your status

The primary thing you live for on Facebook, besides having more friends than anyone else, is to update your status. It's the best way to share your thoughts with the universe, far cheaper than skywriting and far less offensive than a robocall.

To set your status, follow these steps in the Facebook app:

1. **Switch to the News Feed.**

   Tap the News Feed icon, shown in Figure 11-1.

2. **Tap the status update area — specifically, the text *What's On Your Mind*.**

   You see the Post to Facebook screen, where you can type your musings, similar to what's shown in Figure 11-2.

3. **Tap the action bar to choose where to send the new post.**

   Your choices are the news feed (Post to Facebook), a friend's timeline, or a specific group.

4. **Tap the To field.**

   Depending on where you're posting, your choices vary. For posting to Facebook (the news feed), you can choose Public or Friends, where Public makes the post visible to anyone on Facebook and where Friends limits viewing to only your friends. Otherwise, choose a user or group for sharing the post.

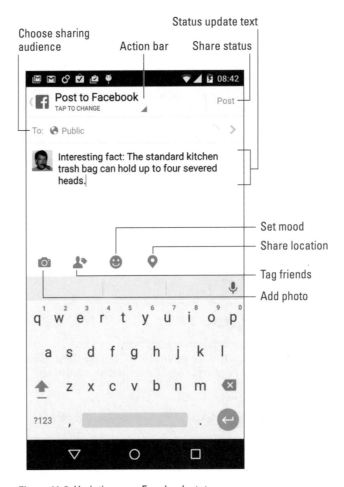

Choose sharing audience

Status update text

Action bar

Share status

Set mood

Share location

Tag friends

Add photo

Figure 11-2: Updating your Facebook status.

5. Tap the *What's On Your Mind* text to type something pithy, newsworthy, or typical of the stuff you read on Facebook.

6. Tap the Post button to share your thoughts.

To cancel the post, tap the Back navigation icon. Tap the Discard button to confirm.

You can also set your status by using the Facebook widget on the Home page, if it's been installed: Tap the What's on Your Mind text box, type your important news tidbit, and tap the Share button.

Uploading a picture to Facebook

One of the many things your Android phone can do is take pictures. Combine that feature with the Facebook app, and you have an all-in-one gizmo designed for sharing the various intimate and private moments of your life with the ogling Internet throngs.

The picture-posting process starts by tapping the Photo icon in the Facebook app. Refer to Figures 11-1 and 11-2 for popular Photo icon locations on the main screen and the Post to Facebook screen. After you tap the Photo icon, the photo selection screen appears. You have two choices:

✔ **First, you can select an image from pictures shown on the screen.** These are images found on the phone. Tap an image, or tap several images to select a bunch, and then tap the Done button.

✔ **Second, you can take a picture by using the phone's camera.** Tap the Add Photo icon, as shown in the margin, to begin the process.

If you elect to use the phone's camera to take a picture, you're switched to the Camera app. Snap the photo. You see a review screen, similar to what's shown in Figure 11-3, although not every phone uses the same Camera app.

Tap the Retry button to take another image, or tap Done / OK and get ready to post the image to Facebook. Tap Cancel to abandon your efforts.

After you select the image, it appears on a new post screen. Continue to create the post as described earlier in this chapter. Tap the Post button to present your efforts to the Facebook community. The image can be found as part of your status update or news feed, and it's also saved to your Mobile Uploads album on Facebook.

✔ See Chapter 13 for more information on using the phone's camera.

✔ I find it easier to take a bunch of images by using the Camera app and then choose an image later to upload it to Facebook.

Retry Done/OK Cancel

Figure 11-3: Snapping a pic for Facebook.

✔ To share something on Facebook when you use another app, tap the Share icon. Choose the Facebook app to share what you're viewing: an image, music, YouTube video, and so on.

Changing the Facebook ringtone

To change or eliminate the noise the Facebook app makes for notifications, follow these steps:

1. **Tap the More icon.**

   Refer to Figure 11-1 for its location.

2. **Choose App Settings.**

3. **Choose Notification Ringtone.**

4. **If prompted with a Complete Action Using card, select Media Storage and tap the Just Once button.**

   See Chapter 19 for information on dealing with the Complete Action Using card, which falls under the topic of "default apps."

5. **Choose a sound from the list.**

   You hear a sound preview, but nothing is set until you:

6. **Tap OK.**

To mute Facebook notifications, choose the sound None in Step 5.

To use the phone's standard notification ringtone, choose Default Notification Sound in Step 5.

Also see Chapter 20 for information on setting the phone's ringtone.

A Virtual Hangout

Google has made various forays into the social networking arena, the most obvious of which is the Google+ service. Though the Google+ app can be found on your phone, a better and more useful social app is Google Hangouts.

The Hangouts app offers text, voice, and video chat. On some phones, Hangouts serves as the text messaging app; see Chapter 8. The Hangouts app can even be used to place phone calls, as long as the companion app, Hangouts Dialer, is installed.

Using Hangouts

Start your Google Hangouts journey by opening the Hangouts app. You may find a launcher on the Home screen, the app might dwell in a Google folder or, like all apps on the phone, it can be found in the apps drawer.

When you first start the Hangouts app, it may ask whether you want to make phone calls. Of course you do! Install the Hangouts Dialer app. If you're not prompted, get that app from the Play Store.

Hangouts hooks into your Google account. Tap the Contacts tab on the left side of the screen, shown in Figure 11-4, to view your Google contacts. Tap the center tab, Conversations, to view current and previous conversations. When the Hangouts Dialer is installed, a third tab, Dialer, appears on the right side of the screen.

Figure 11-4: Google Hangouts.

You can do anything else on the phone while the Hangouts app stays active. You're alerted via notification of an impending Hangouts request. The notification icon is shown in the margin.

- ✔ Conversations are archived in the Hangouts app. To peruse a previous text chat, tap the Conversations tab and choose a conversation from the list. Video calls aren't archived, but you can review when the call took place and with whom.

- ✔ If you use Hangouts as the phone's text messaging app, you'll find the messages archived on the Conversations tab.

- ✔ Video calls aren't archived, but you can review when the call took place and with whom by choosing a video chat item.

- ✔ To remove a previous conversation, long-press it. Tap the Trash icon that appears atop the screen.

- ✔ When installed, the Hangouts Dialer app lets you place phone calls from within the Hangouts app. Tap the Dialer tap, and then type a contact name or phone number. Tap the matching contact or phone number to dial.

- ✔ To use Hangouts, your friends must have a Google account. They can be using a computer or mobile device; it doesn't matter which. But they must have a camera available to enable video chat.

Typing at your friends

Text chatting is one of the oldest forms of communication on the Internet. People type text back and forth at each other, which can be tedious, but it remains popular. To text-chat in the Hangouts app, obey these steps:

1. **Tap the Contacts tab, or to continue a previous conversation, tap the Previous Conversations tab.**

2. **Choose a contact or previous conversation.**

3. **Use the onscreen keyboard to type a message, as shown in Figure 11-5.**

4. **Tap the Send icon to send your comment.**

   The Send icon replaces the Attachment icon when you start to type.

You type, your friend types, and so on until you grow tired or the phone's battery dies.

- ✔ To add more people to the hangout, tap the Action Overflow and choose New Group Conversation. Choose a friend from those listed to invite him into the hangout.

- ✔ Tap the Attachment icon to add a photo to the conversion, such as the image shown in Figure 11-5. The Attachment icon, shown in the margin, appears in the Send Hangouts Message text box, but only when you haven't yet typed anything.

Type here   Voice chat

Conversation   Video chat

Figure 11-5: Text-chatting.

Talking and video chat

Take the hangout up a notch by tapping the Voice Chat icon (refer to Figure 11-5) When you do, your friend receives a pop-up invite. After that person taps the Accept icon, you begin talking.

To see the other person, tap the Video Chat icon. As soon as they agree, you can see each other, as shown in Figure 11-6. The person you're talking with appears in the big window; you're in the smaller window, as shown in the leftmost figure.

To mute the call, tap the Microphone icon; tap the Video icon (it's on the lower right of the right side of Figure 11-6) to return to voice or text chat. The onscreen controls may vanish after a second; tap the screen to see the controls again.

To end the conversation, tap the End Call icon, at the bottom center. Well, say goodbye first, and then tap the icon.

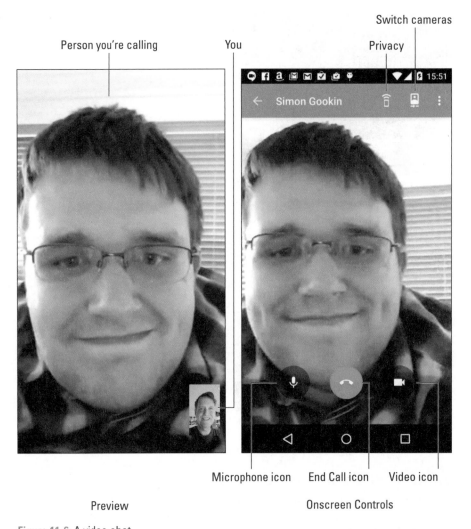

Figure 11-6: A video chat.

Let's All Tweet

Twitter is a social networking site that lets you share short bursts of text, or *tweets*. You can create your own or just choose to follow others, including organizations, governments, celebrities, and/creatures from alien planets.

If your phone didn't come with the Twitter app, obtain it from the Play Store, as described in Chapter 16. Install and run the app to sign in to Twitter using an existing account, or create a new account on the spot.

Figure 11-7 shows the Twitter app's main screen, which shows the current tweet feed. The Twitter app is updated frequently, so its exact appearance may change after this book has gone to press.

Figure 11-7: The Twitter app.

To read tweets, choose the Home category, shown in Figure 11-7. Recent tweets are displayed in a list, with the most recent information at the top. Scroll the list by swiping it with your finger. Drag the list downward to refresh.

To tweet, tap the New Tweet icon, shown in Figure 11-7. The newtweet screen appears, where you can compose your tweet.

A tweet can be no more than 140 characters long. That number includes spaces and punctuation. For direct messages, Twitter has raised the limit above 140 characters.

Tap the Tweet button to share your thoughts with the twitterverse.

> ✔ The Twitter app comes with companion widgets that you can affix to the Home screen. Use the widgets to peruse recent tweets or compose a new tweet. Refer to Chapter 19 for information on affixing widgets to the Home screen.

✔ You can share material from your phone with Twitter by tapping the Share icon in various apps. The time you're looking at is then added to a new tweet.

Skype the World

The Skype service is used the world over as a free way to make Internet phone calls and to video-chat. Plus, if you're willing to pony up some money, you can make inexpensive calls to phones around the world.

The typical Android phone doesn't come with the Skype app preinstalled. To get Skype, visit the Play Store and obtain the Skype app.

To use Skype, you need a Skype account. You can sign up when you first open the app, or you can visit www.skype.com on a computer and complete the process by using a nice, full-size keyboard and widescreen monitor.

Skype's popular features include text chat as well as voice and video chat, although you can use these features only with fellow Skype users. The big enchilada in Skype is placing phone calls. Because you already have a phone, that feature's most important aspect is placing inexpensive international calls. Here's how it works:

1. **Tap the Dialpad icon on the Skype app's Home screen.**

   See Figure 11-8 for its appearance and location, although the Skype app is frequently updated and things will change in the future.

2. **Use the keypad to type the international number.**

   The number begins with a plus sign (+) followed by the county code, and then the phone number. You can choose a country by tapping the screen as shown in Figure 11-8.

3. **Tap the Dial icon to dial the number.**

4. **When the other party answers, talk.**

These steps meet with success only when you've purchased Skype Credit. If not, tap your account icon (see Figure 11-8) and follow the onscreen steps to add credit. Don't worry! You don't need a lot of Skype Credit to place an international call.

✔ Text, voice, and video chat on Skype over the Internet are free. When you use a Wi-Fi connection, you can chat without consuming your cellular plan's data minutes.

✔ If you plan to use Skype a lot, get a good headset.

Account icon   Select County Code   Backspace/Delete

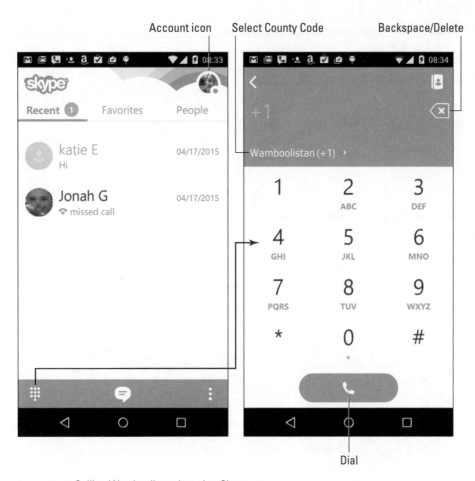

Dial

Figure 11-8: Calling Wamboolistan by using Skype.

Part III
Amazing Phone Feats

Learn how to save a chunk of the map for offline viewing by visiting www.dummies.com/extras/androidphones.

In this part . . .

✓ Understand how to use the Maps app.

✓ Work with the phone's camera.

✓ Explore music on an Android phone.

✓ Discover interesting apps.

✓ Get more apps, music, eBooks, and movies.

12

There's a Map for That

In This Chapter

▶ Using the Maps app
▶ Adding layers to the map
▶ Finding your location
▶ Sharing your location
▶ Searching for places
▶ Saving favorite places
▶ Getting directions and navigation

*W*here are you? More importantly, where is the nearest Mexican restaurant? Normally, these questions can be best answered when you're somewhere familiar or with someone who knows the territory or while standing at a brightly lit intersection and the aroma of chili peppers and tequila is wafting from a restaurant nearby and the sign says *El Viejo Buey*.

The answer to the questions "Where are you?" and "What's nearby?" are readily provided by the Maps app. This app uses your phone's global positioning system (GPS) capabilities to gather information about your location. It also uses Google's vast database to locate interesting places nearby and even get you to those locations. The Maps app is truly amazing and useful — after you understand how it all works.

Basic Map

Perhaps the best thing about using the Maps app is that there's no risk of improperly folding anything. Even better, the Maps app charts everything you need to find your way: freeways, highways, roads, streets, avenues, drives, bike paths, addresses, businesses, and points of interest.

Using the Maps app

Start the Maps app by tapping its launcher on the Home screen, or you can summon the app by locating it in the apps drawer. If you're starting the app for the first time or it has recently been updated, you must agree to the terms and conditions. Do so as directed on the touchscreen.

Your Android phone communicates with GPS satellites to hone in on your current location. (See the later sidebar "Activate your location!") That location appears on the map, similar to Figure 12-1. The position is accurate to within a given range, as shown by a blue circle around your location. If the circle doesn't appear, your location is either pretty darn accurate or you need to zoom in.

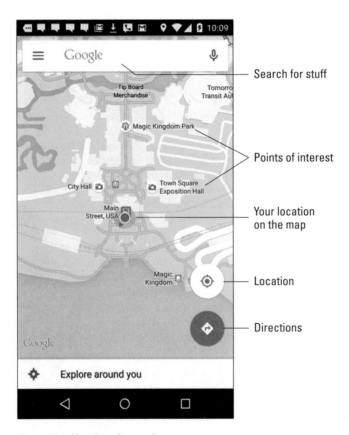

Figure 12-1: Your location on the map.

Here are some fun things you can do when viewing the basic street map:

Zoom in: To make the map larger (to move it closer), double-tap the screen. You can also spread your fingers on the touchscreen to zoom in.

Zoom out: To make the map smaller (to see more), pinch your fingers on the touchscreen.

Pan and scroll: To see what's to the left or right or at the top or bottom of the map, swipe your finger on the touchscreen. The map scrolls in the direction you swipe.

Rotate: Using two fingers, rotate the map clockwise or counterclockwise. Tap the Compass Pointer icon, as shown in the margin, to reorient the map with north at the top of the screen.

Perspective: Touch the screen with two fingers and swipe up or down to view the map in perspective. You can also tap the Location icon to switch to perspective view, although that trick works only for your current location. To return to flat-map view, tap the Compass Pointer icon.

The closer you zoom in to the map, the more detail you see, such as street names, address block numbers, and businesses and other sites — but no tiny people.

✔ The blue triangle (refer to Figure 12-1) shows in which general direction the phone is pointing.

✔ When the phone's direction is unavailable, you see a blue dot as your location on the map.

✔ When the location icon is blue, you're viewing your current location on the map. Tap the icon to enter perspective view. Tap the Perspective icon, shown in the margin, to return to flat-map view.

✔ When all you want is a virtual compass, similar to the one you lost as a kid, get a compass app from the Google Play Store. See Chapter 16 for more information about the Google Play Store.

Adding layers

You add details to the map by applying layers: A *layer* enhances the map's visual appearance by providing more information, or by adding other fun features to the basic street map, such as the Satellite layer, shown in Figure 12-2.

Activate your location!

The Maps app works best when you activate all the phone's location technologies. Open the Settings app and choose the Location item. On some Samsung phones, the Location item is found on the Connections tab. Ensure that the master control is in the On position.

Your Android phone uses several technologies to hone in on your location, but only when the Location setting is activated. It also uses the phone's cellular signal as well as Wi-Fi. In fact, if Wi-Fi is disabled, you may be prompted to activate it to help assist the phone in pinpointing your location. See Chapter 17 for information on Wi-Fi.

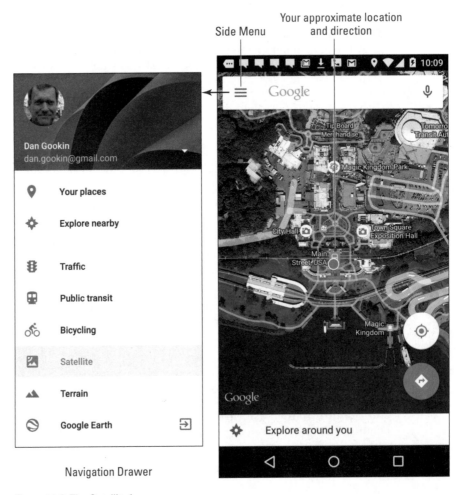

Side Menu

Your approximate location and direction

Navigation Drawer

Figure 12-2: The Satellite layer.

The key to accessing layers is to tap the Side Menu icon to view the navigation drawer. It displays several layers you can add to the map, such as the Satellite layer, shown in Figure 12-2. Another popular layer is Traffic, which lists updated travel conditions.

To remove a layer, choose it again from the navigation drawer; any active layer appears highlighted. When a layer isn't applied, the street view appears.

It Knows Where You Are

Many war movies have this cliché scene: Some soldiers are looking at a map. They wonder where they are, when one of them says, "We're not even on the map!" Such things never happen on your phone's Maps app. That's because it always knows where you are.

Well, unless you're on the planet Venus. I've heard that the Maps app won't work there.

Finding a location

The Maps app shows your location as a blue dot on the screen. But *where* is that? I mean, if you need to phone a tow truck, you can't just say, "I'm the blue dot on the orange slab by the green thing."

Well, you *can* say that, but it probably won't do any good.

To view your current location, tap the Location icon, as shown in the margin. If you desire more information, or if you want details about any random place, long-press the Maps screen. Up pops a card, similar to the one shown in Figure 12-3. The card gives your approximate address.

Tap the card to see a screen with more details and additional information, as shown on the right in Figure 12-3.

✔ This trick works only when Internet access is available and the Maps app can communicate with the Google map servers.

✔ To make the card go away, tap anywhere else on the map.

✔ For nearby locations, the Route icon (see Figure 12-3) shows a method of travel and an approximate time value for the trip duration. Also see the later section "Your Phone Is Your Copilot."

✔ When you have way too much time on your hands, play with the Street View feature. Tap that item on the location's card to examine a location preview with a 360-degree perspective. It's a great way to familiarize yourself with a destination or to plan a burglary.

Long-press a location
to see the address

Mark the location
as a favorite

Share this location

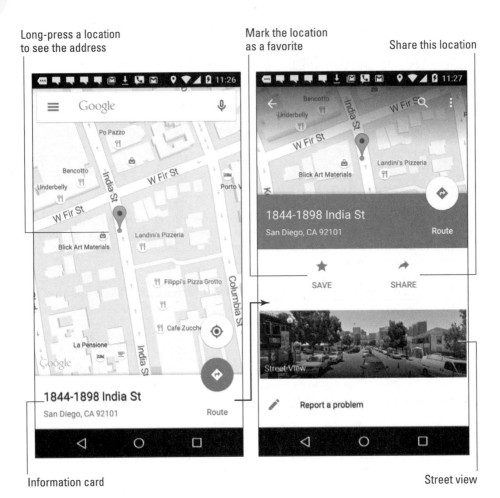

Information card

Street view

Figure 12-3: Finding an address.

Helping others find your location

You can use your Android phone to send your current location to a friend. If your pal has a phone with smarts similar to those of your Android phone, he can use the coordinates to get directions to your location. Maybe he'll even bring tacos!

To send your current location to someone else, obey these steps:

1. **Long-press your current location on the map.**

   To see your current location, tap the Location icon in the lower right corner of the Maps app screen.

   After long-pressing your location (or any location), you see a card displayed showing the approximate address.

2. **Tap the card, and then tap the Share icon.**

   Refer to Figure 12-3 for this icon's location.

3. **Choose the app to share the location.**

   For example, choose the phone's text messaging app, Gmail, Email, or whichever useful app is listed.

4. **Use the selected app to complete the process of sending your location to someone else.**

As an example, you can send your location by using the phone's text messaging app. The recipient can tap the link in the text message to open your location in his phone's Maps app. When the location appears, the recipient can follow my advice in the later section "Your Phone Is Your Copilot" to reach your location. Don't loan him this book, either — have him purchase his own copy. And bring tacos. Thanks.

Find Things

The Maps app can help you find places in the real world, just as the Google Search app helps you find places on the Internet. Both operations work basically the same way: Open the Maps app and type something to find into the Search text box (refer to Figure 12-1). You can type a variety of terms in the Search box, as explained in this section.

Looking for a specific address

To locate an address, type it in the Search box; for example:

```
1600 Pennsylvania Ave., Washington, D.C. 20006
```

As you type, suggestions appear onscreen. Tap a matching suggestion to view that location. Otherwise, tap the onscreen keyboard's Search key, and that location is shown on the map.

After finding a specific address, the next step is getting directions. See the later section "Your Phone Is Your Copilot."

- ✔ If you omit the city name or Zip code, the Maps app looks for the closest matching address near your current location.

- ✔ Tap the X button in the Search box to clear the previous search.

Finding a business, restaurant, or point of interest

You may not know an address, but you know when you crave sushi or perhaps the exotic flavors of Kamchatka. Maybe you need a hotel or a gas station, or you have to find a place that patches toupees. To locate a business entity or a point of interest, type its name in the Search box; for example:

```
movie theater
```

This search text locates movie theaters on the current Maps screen. Or, to find locations near you, first tap the Location icon (shown in the margin) and then type the search text.

To look for points of interest at a specific location, add the city name, district, or Zip code to the search text. For example:

```
sushi 98109
```

After typing this command and tapping the Search button, you see a smattering of sushi bars found near downtown Seattle, similar to those shown on the left side of side of Figure 12-4.

To see more information about a result, tap its card, such as the one for I Love Sushi on Lake Union in Figure 12-4. To get to the location, tap the Route icon on the location's card, as illustrated in the figure. See the later section "Your Phone Is Your Copilot."

- ✔ Every dot on the screen represents a search result (see Figure 12-4).

- ✔ Spread your fingers on the touchscreen to zoom in to the map.

- ✔ The search results card (on the right in Figure 12-4) lists business hours as well as other information. Tap the Call icon to place a call. Tap the Website icon to visit the location's website, if available.

- ✔ If you *really* like the location, tap the Save (Star) icon. That directs the Maps app to keep the location as one of your favorite places. The location is flagged with a Star icon on the Maps app screen. See the next section.

- ✔ When you can't make up your mind or you just want to see what's out there, tap the Location icon and choose the item Explore Around You, found at the bottom of the screen (refer to Figure 12-1). Peruse the list of locations to find something interesting.

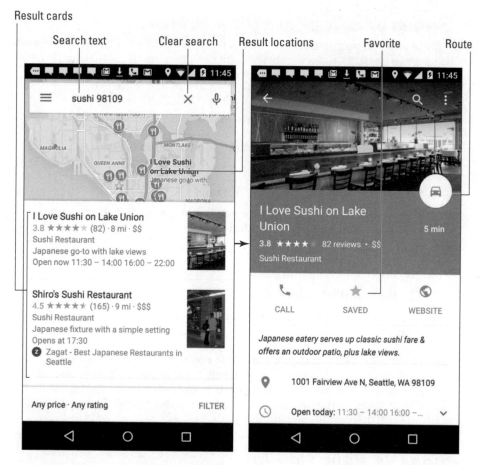

Figure 12-4: Search results for sushi in Seattle.

Searching for favorite or recent places

Just as you can bookmark favorite websites on the Internet, you can mark favorite places in the real world by using the Maps app. The feature is called Your Places.

To review your favorite places or browse your recent map searches, tap the Side Menu icon and choose Your Places from the navigation drawer. Swipe through the list to see recent searches, saved places, and any offline maps you've saved. To revisit a place, tap its entry in the list.

Mark a location as a favorite by tapping the Save (Star) icon on the location's details card.

Setting your Home and Work locations

Two places you visit most frequently in the real world are where you live and where you work. Two favorite places you can create in the Maps app provide quick access to those special locations. They're called, logically enough, Home and Work.

To set your Home and Work locations, follow these steps in the Maps app:

1. **Tap the Side Menu icon to display the navigation drawer.**

2. **Choose Settings.**

3. **Choose the item Edit Home or Work.**

4. **Choose the Edit Home Address item and type in your home address, or the location you prefer as "home."**

   As you type, choose a matching location from those listed on the screen.

5. **Tap the Back navigation icon.**

6. **Choose Edit Work Address.**

7. **Type the address for your office or wherever you work or pretend to work.**

You can use the Home and Work shortcuts when searching for a location or getting directions. For example, type **Home** into the Search box to instantly see your home location. To get directions to or from your current location to work, type **Work** as the destination or starting point, respectively.

Your Phone Is Your Copilot

One command that's associated with locations found in the Maps app deals with getting directions. The command is called Route, and it shows either the Route icon (see the margin) or a mode of transportation, such as a car, bike, bus, or ornithopter. Here's how it works:

1. **Tap the Route icon on a location's card.**

   You see a route card, similar to the one shown on the left in Figure 12-5. It lists the name of the destination, which is the I Love Sushi restaurant in the figure.

2. **Choose a method of transportation.**

   The available options vary, depending on your location. In Figure 12-5, the items are (from left to right) Car, Public Transportation, On Foot, and Bicycle.

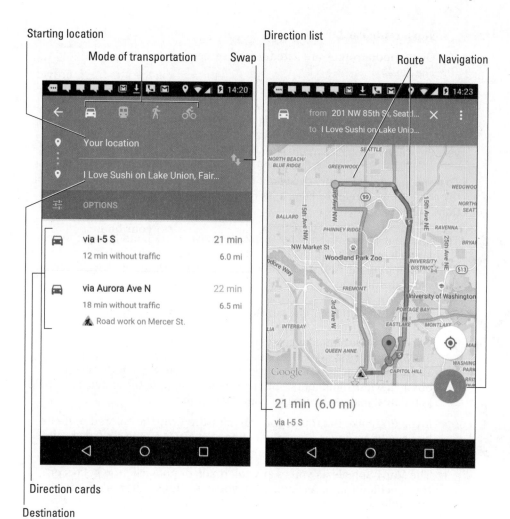

Figure 12-5: Planning a trip.

3. Set a starting point.

The starting point is listed as Your Location, which is the phone's current location. You can type in another location or use the Home or Work shortcuts, as described earlier in this chapter.

4. Ensure that the starting location and destination are what you want.

If they're reversed, tap the Swap icon (refer to Figure 12-5).

5. **Tap a route card.**

One or more routes are listed on the screen, as shown on the right in Figure 12-5.

6. **Peruse the results.**

The map shows your route, highlighted as a blue line on the screen. Detailed directions also appear. Traffic jams show up as red, with slow traffic as yellow, which is coincidentally shown on the right in Figure 12-5.

To see a list of directions, check out the directions list card at the bottom of the route screen (refer to Figure 12-5). A scrolling list of turn-by-turn directions appears.

To begin your journey and enter turn-by-turn Navigation mode, tap the Navigation icon, as shown in Figure 12-5 and in the margin. You can mute the voice by tapping the Action Overflow and choosing Mute. Otherwise, toodle on to your destination.

To exit from Navigation mode, tap the Close icon on the screen.

✔ Two things are a must when you use your phone to navigate while driving your car. First, plug the phone into a power source, such as the car's 12-volt power supply, formerly known as a cigarette lighter. Second, obtain a windshield-mount phone cradle. Together, these items can help you quickly and safely get to your destination.

✔ If you don't like the route, you can adjust it: Drag the colored route line by using your finger. Time and distance measurements shown on the cards change as you adjust the route.

✔ The Maps app alerts you to any toll roads or construction delays on the specified route. As you travel, you can choose alternative routes, if available.

✔ You may not get perfect directions from the Maps app, but it's a useful tool for places you've never visited.

✔ The phone stays in Navigation mode until you exit. A navigation notification can be seen atop the touchscreen while you're in Navigation mode.

13

Pics and Vids

The second thing that Alexander Graham Bell thought of, right after the busy signal, was the camera phone. Sadly, it proved somewhat impractical because cameras and phones of the 1890s weren't really portable. And it was difficult to talk on the phone and hold still for 60 seconds to get the best exposure.

Not until the end of the 20th century did the marriage of the cell phone and digital camera become successful. It may seem like an odd combination, yet it's quite handy to have the phone act as a camera because most people keep their phones with them. That way, should you chance upon Bigfoot or a UFO, you can always take a picture or video — even when there's no cell signal, which is often the case when I try to photograph Bigfoot.

The Phone's Camera

A camera snob will tell you that no true camera has a ringtone. You know what? He's correct: Phones don't make the best cameras. Regardless, your phone has a camera. It can capture both still and moving images. That task is carried out by using the Camera app.

- The Camera app features two basic modes: Still Shot and Video. The same app handles both shooting modes.

- Many camera apps feature additional shooting modes, such as Panorama. These shooting modes are chosen in the same way as the basic Still Shot and Video modes are chosen, which differs between the various incarnations of the Camera app.

- Some Camera apps allow you to add special effects to your images, whereas other apps are sparse on effects. The Photos app, covered later in this chapter, also lets you add image effects.

- This section covers using the Google Camera app. Your phone most likely uses a similar, though not identical, app. Many of the directions in this section may not be specific to your phone's Camera app.

- When you use the Camera app, the navigation icons (Back, Home, Recent) turn into tiny dots or vanish altogether. The icons are still there, and they still work; tap or swipe the screen to see them.

Using basic camera controls

Here are some pointers that apply to all variations of the Camera app:

- Use the phone in either landscape or portrait orientation while taking a still shot. Don't worry either way: The image is always saved with the proper side up.

- I strongly recommend that you record video in horizontal orientation only. This presentation appears more natural.

- The phone's touchscreen serves as the viewfinder; what you see on the screen is exactly what appears in the final photo or video.

- Tap the screen to focus on a specific object. You'll see a focus ring or square that confirms how the camera lens is focusing. Not every phone's camera hardware can focus; the front-facing camera features a fixed focus.

- Zoom in by spreading your fingers on the screen.

- Zoom out by pinching your fingers on the screen.

- Some phones let you use the Volume key to zoom in or out: Up volume zooms in, Down volume zooms out. If the Volume key doesn't zoom, it may serve as a second shutter button (to snap the picture). While you're shooting video, the Volume key may serve to snap a still shot.

- Hold the phone steady! I recommend using two hands for both taking a still shot and shooting video.

- The phone stores pictures and videos in the DCIM/Camera folder. Still images are stored in the JPEG or PNG file format; video is stored in the MPET-4 format. If your phone offers removable storage, the Camera app automatically saves images and videos to that location.

Snapping a still image

Taking a still image requires only two steps: First, ensure that the Camera app is in Single Shot mode; second, tap the Shutter icon to snap the photo.

 Set Single Shot mode in the Google Camera app by swiping inward from the left side of the screen, as illustrated in Figure 13-1. Tap the Single Shot icon, shown in the margin. Coincidentally, the Shutter icon also looks like the Single Shot icon. Weird.

Camera modes (swipe in from left)

Shutter icon/current mode

Navigation icons (hidden)

Camera settings

Figure 13-1: The Google Camera app.

In other Camera apps, choose the icon representing Single Shot mode. Sometimes, you'll find a toggle that switches between Single Shot and Video modes. If so, ensure that the setting is for Single Shot to take a still image.

Set the resolution *before* you shoot. See the later section "Setting the resolution."

Recording video

To capture video, switch the Camera app to Video Recording mode. The Shutter icon becomes the Record icon, similar to the one shown in the margin. Tap that icon to start recording. The elapsed time and maybe even the amount of storage consumed appear on the touchscreen as video is being recorded, as shown in Figure 13-2. Tap the Stop icon to end the recording.

Figure 13-2: Recording video on a Samsung phone's Camera app.

In the stock Android Camera app, choose the video mode as shown in Figure 13-1. For other camera apps, such as the Camera app shown in Figure 13-2, tap the Video Mode icon to either switch modes to video or just start recording.

✔ Some Camera apps feature a Pause icon for recording video. Tap that icon to temporarily suspend recording. Tap the Record icon to resume. The video you record is saved as a single video as opposed to multiple videos.

✔ Some versions of the Camera app may allow you to grab a still image while the phone is recording: Tap the screen. The image is snapped and saved.

Deleting immediately after you shoot

Sometimes, you just can't wait to banish an image to bit hell. Either an annoyed person is standing next to you, begging that the photo be deleted, or you're just not happy and you feel the urge to smash into digital shards the picture you just took. Hastily follow these steps:

1. **Tap the image thumbnail that appears on the screen, or swipe the screen to the left.**

   Either technique displays the image you just snapped. If you don't see the thumbnail, tap the screen. If you still don't see it, swipe the screen from right to left.

2. **Tap the Delete icon.**

   If you don't see the icon, tap the screen so that the icon shows up.

3. **If prompted, tap the OK or Delete button to confirm.**

See the later section "Your Digital Photo Album" for more information on managing the phone's images.

Setting the flash

The Camera app features three flash settings, illustrated in Table 13-1. The current setting may be shown on the screen, or you may have to access the flash setting control to view the current setting.

Table 13-1		Camera Flash Settings
Setting	*Icon*	*When the Flash Activates*
Auto	⚡A	During low-light situations
On	⚡	Always
Off	⚡̸	Never

To change or check the flash setting in the Google Camera app, tap the Action Overflow on the app's main screen. The current flash setting is represented as an icon similar to those referenced in Table 13-1. Tap that icon to cycle through and set the camera's flash setting.

Some variations of the Camera app show the flash setting on the Camera app's main screen. If so, tap that icon to cycle through the flash settings.

✔ A good time to turn on the auto flash is when taking pictures of people or objects in front of something bright, such as a lovely calico kitten purring in front of an exploding volcano.

✔ A "flash" setting is also available for shooting video in low-light situations. In that case, the flash LED is on the entire time. This setting is made similarly to setting the flash for still images, although the options are only On and Off. It must be set before you shoot video, and, yes, it devours a lot of battery power.

Setting the resolution

You don't always have to set the highest resolution or top quality for images and videos you shoot by using your Android phone. Especially when you're shooting for the web or uploading pictures to Facebook, top quality is a waste of storage space because the image is shown on a relatively low-resolution computer monitor.

As you may suspect, setting the image resolution or video quality is done differently in each phone's Camera app. The Google Camera app provides access to these settings by following these steps:

1. **Display the Camera app's shooting modes.**

   Swipe the screen from left to right to view shooting modes, as shown earlier in Figure 13-1.

2. **Tap the Settings icon.**

3. **Choose Resolution & Quality.**

   The Resolution & Quality screen is organized by shooting mode: Camera (still image) and Video. It's further organized by back or front camera.

4. **Choose a mode and a camera.**

   For example, tap Back Camera Photo to set the still-image resolution for the phone's rear camera.

5. **Choose a resolution or video quality setting from the list.**

   Options are presented in aspect ratio as well as in megapixels.

For other Camera apps, the still-shot resolution or video quality items may be found by tapping an Action Overflow or Settings icon and then choosing a resolution, picture size, or similar action. You may also have to switch cameras (front or rear) before setting the resolution for each camera.

✔ Set the resolution or video quality *before* you shoot! Especially when you know where the video will end up (on the Internet, on a TV, or in an email), it helps to set the quality first.

✔ The resolution and video quality choices are more limited on the front-facing camera because it's not as sophisticated as the rear camera.

✔ A picture's *resolution* describes how many pixels, or dots, are in the image. The more dots, the better the image looks and prints.

✔ An aspect ratio describes the image's overall dimensions horizontally by vertically. A 4:3 ratio is pretty much standard, just like the phone's screen. The 16:9 ratio is the widescreen format.

✔ The video quality settings HD and SD refer to High Definition and Standard Definition, respectively. The "p" value represents vertical resolution, with higher values indicating higher quality.

✔ *Megapixel* is a measurement of the amount of information stored in an image. A megapixel is approximately 1 million pixels, or individual dots that compose an image. It's often abbreviated MP.

Shooting yourself

Providing that your phone features a front-facing camera, you can direct the Camera app to use that camera to shoot still images and record videos of yourself. The vernacular term is *selfie* for self-shot. Get ready with your duck face!

Use the Camera app to switch between front and rear cameras. On the stock Android Camera app, tap the Action Overflow and then tap the Switch Cameras icon, similar to the one shown in the margin. The icon might also appear as a camera with arrows pointing both left and right.

When you see yourself in the Camera app's viewfinder, you've completed the task successfully. Smile. Click.

Tap the same icon again to switch back to the rear camera. The icon may change its appearance, but you should find it in the same location in the Camera app.

Your Digital Photo Album

The traditional Android photo-management/album app is Gallery. Google is slowly replacing it with the Photos app. Your phone may sport one or both apps, which work similarly. The Photos app offers a few more features, plus better integration with online photo sharing, so it's the photo management app covered in this chapter.

Viewing your photos and videos

The Photos app organizes your photos and videos in several ways. The Photos screen, shown in Figure 13-3, lists photos by date. You can choose more specific views by choosing other items from the navigation drawer, also shown in the figure.

Tap an image to view it full-screen, as shown on the right in Figure 13-3. You can then swipe the screen left or right to peruse other images.

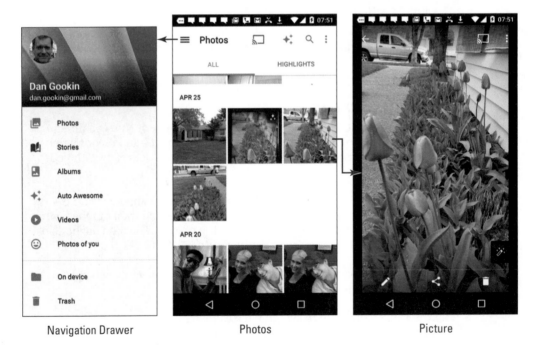

Navigation Drawer   Photos   Picture

Figure 13-3: Image organization in the Photos app.

Videos stored in an album appear with the Play icon. Tap that icon to play the video. As the video is playing, tap the screen again to view onscreen controls.

> ✔ While viewing an image or video full-screen, the navigation icons may disappear. Tap the screen to view them.
>
> ✔ Tap the Back navigation icon to return to an album after viewing an image or video.

Starting a slideshow

The Photos app can display a slideshow of your images, but without the darkened room and sheet hanging over the mantle. To view a slideshow, first view an image full-screen. Tap the Action Overflow and choose Slideshow. Images from that particular album or date appear one after the other on the screen.

Tap the Back navigation icon to exit the slideshow.

Slideshows don't have to remain on your phone's diminutive screen: If a nearby HDMI TV or monitor features a Chromecast dongle, tap the Chromecast icon, as shown in the margin. Choose a specific Chromecast device from the list to view the slideshow on the big screen.

To end a slideshow in Chromecast, tap the Chromecast icon again and tap the Disconnect button. See Chapter 18 for more information on screencasting.

Finding a picture's location

In addition to snapping a picture, your phone also records the specific spot on Planet Earth where the picture was taken. This feature is called Location Tags, GPS-tag, or Geo-Tag.

To view location-tag information in the Photos app, heed these directions:

1. **View the image.**

2. **Tap the Action Overflow and choose Details.**

   You see a card displaying image details, similar to what's shown in Figure 13-4. Map information, if available, appears atop the card.

Alas, the map preview is just that: It's not interactive, and you can't instantly hop to the Maps app to browse around — although that option might be available on some phones.

The Location Tags feature isn't without controversy. People either love it or hate it. If you hate it, you can disable the feature by heeding these directions in the Google Camera app:

1. **View the Camera modes.**

   Swipe the screen from the left edge toward the center to view the shooting modes (refer to Figure 13-1).

2. **Tap the Settings icon.**

   In some Camera apps, the Settings icon is available without your having to display the shooting modes.

Map preview

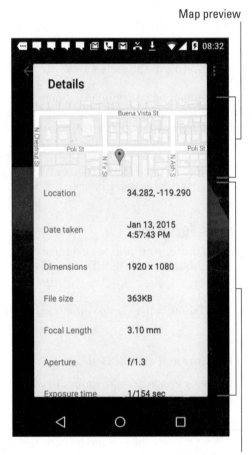

Image trivia and such

Figure 13-4: Image details, including location.

3. **Enable the Save Location or Location Tags feature.**

   Use the master control to enable the feature — or disable it if you don't want location information saved with your photos.

Other Camera apps may require you to tap an Action Overflow or Settings icon to find the location-tag item. You may further have to tap a Settings tab to find this specific item.

   ✏ Deactivating the Location Tags feature doesn't remove that information from photos you've already taken.

   ✏ The location-tag information is stored in the picture itself. Other devices, apps, and computer programs can then read this information to determine where the image was taken.

Sharing your images

Just about every app wants to get in on the sharing bit, especially when it comes to pictures and videos. The key is to view an item in the Photos app and then tap the Share icon, as shown in the margin. Choose an app to share the image or video, and that item is instantly sent to that app.

What happens next?

That depends on the app. For Facebook, Twitter, and other social networking apps, the item is attached to a new post. For Gmail, the item becomes a message attachment. Other apps treat images and videos in a similar manner, somehow incorporating the item(s) into whatever wonderful thing that the app does. The key is to look for that Share icon.

Posting a video to YouTube

The best way to share a video is to upload it to YouTube. As a Google account holder, you also have a YouTube account. Why not populate that account with your latest, crazy videos? Who knows what may go viral next!

To upload a video you've recorded, follow these steps:

1. **Ensure that the Wi-Fi connection is activated.**

   The best way to upload a video is to use the Wi-Fi connection, which won't incur data surcharges. In fact, if you do use the 4G LTE network for uploading a YouTube video, you see a suitable reminder about the data surcharges.

2. **Open the Photos app.**

3. **View the video you want to upload.**

   You do not need to play the video. Just have it on the screen.

4. **Tap the Share icon.**

   If you don't see the Share icon, tap the screen.

5. **Choose YouTube.**

   The Upload Video screen appears. You may first see a tutorial on trimming the video, which is the next step.

6. **Trim the video, resetting the starting and ending points.**

   This video-editing step is optional. If you opt to trim, adjust the starting and ending points for the video by dragging them left or right. As you drag, the video is scrubbed, allowing you to preview the start and end points.

7. **Type the video's title.**

8. **Set other options.**

   Type a description, set the privacy level, add descriptive tags, and so on.

9. **Tap the Send icon.**

   The video is uploaded to your YouTube account, which may take some time to complete.

The Uploading notification appears while the video is being sent to YouTube. Feel free to do other things with your phone while the video uploads. When the upload has completed, the Uploading notification stops animating and becomes the Uploads Finished notification.

To view your video, open the YouTube app. See Chapter 15 for details on using the YouTube app.

Image Management

The Photos app offers tools for basic image management, editing, and even adding special effects. I could write an entire book on this app and probably not cover everything you can do. Rather than pester my publisher, I'll constrain myself and cover some of the more basic image editing options.

Editing an image

To edit an image in the Photos app, you must activate Image Editing mode. This mode also enables some of the app's special effects and image editing features, but it all starts by following these steps:

1. **Display the image you want to edit or otherwise manipulate.**

2. **Tap the Edit icon.**

The Edit icon is shown in the margin. If you don't see it, tap the screen and it shows up.

The Editing screen is shown in Figure 13-5. A scrolling list of tools appears at the bottom of the screen. Swipe the list left or right to view the lot. Choose a tool to manipulate the image.

Tap the Review icon (see Figure 13-5) to compare the original and modified images. When you long-press the icon, the original image is shown.

Figure 13-5: Editing an image in the Photos app.

To set your changes, tap the Done icon, or tap the Cancel icon to discard your edits.

Choosing Done or Cancel doesn't end Editing mode. To complete your edits and save the image, tap the Done icon on the main editing screen (refer to Figure 13-5).

- ✔ To view the original version of an image, view that image in the Photos app and tap the Original/Edited icon, shown in the margin. Tapping the icon doesn't restore the image, but it does temporarily show you the original.

- ✔ To restore an image, removing all edits, edit the image again, tap the Action Overflow, and choose Revert. Tap the Revert button to confirm. Tap the Done button to leave Image Editing mode. All edits are removed.

Cropping an image

To *crop* an image is to snip away parts you don't want or need, such as that guy on the far left who photobombed your family picture. To crop an image in the Photos app, obey these steps:

1. **Summon the image for editing.**

   Refer to the preceding section for step-by-step directions.

2. **Choose the Crop tool.**

3. **Adjust the rectangle on the screen to select which portion of the image to keep.**

   Drag the rectangle around. Drag its edges or sides to resize the rectangle.

4. **Save the cropped image.**

5. **Tap the Done icon to exit Editing mode.**

The cropped image is saved in the Photos app.

You can undo the crop by tapping the Restore icon when viewing the image in the Photos app. Refer to the preceding section.

Rotating pictures

Showing someone else an image on your phone can be frustrating, especially when the image is a vertical picture that refuses to fill the screen when the phone is in a vertical orientation. You can fix that issue by rotating the picture in the Photos app. Follow these steps:

1. **Summon the cockeyed image for editing.**

   Refer to the earlier section "Editing an image" for specific directions.

2. **Choose the Rotate tool.**

3. **Tap the Rotate Left or Rotate Right button to reorient the image in 90-degree increments.**

   These buttons are found in the lower right corner of the editing screen.

4. **Tap the Done icon to save the rotated image.**

5. **Tap the Done icon to exit Image Editing mode.**

It's possible to rotate the image to almost any angle. In Step 3, drag your finger around the touchscreen. The image is automatically resized and cropped to maintain its aspect ratio.

Deleting photos and videos

It's entirely possible, and often desirable, to remove unwanted, embarrassing, or questionably legal images and videos from the Photos app.

To banish something to bit hell, tap the Delete (Trash) icon on the screen when viewing an image or a video. Tap the Delete Everywhere button to confirm. The item is gone.

If you don't see the Trash icon, the item cannot be deleted. It's most likely a copy pulled in from a web photo-sharing service or a social networking site.

To undelete an image, tap the Side Menu icon to display the navigation drawer. Choose Trash. Long-press the item to select it, and then tap the Restore icon, as shown in the margin.

O Sweet Music!

In This Chapter

▶ Listening to a song

▶ Building up a song queue

▶ Adding music to your library

▶ Organizing your music into playlists

▶ Managing playlists

▶ Listening to streaming music

*p*ush-button phones and tone dialing were introduced in the 1970s. People realized that the tones generated by dialing certain numbers sounded like popular tunes. My parents' home number sounded like *Yankee Doodle*. On your Android phone, however, listening to music is much more sophisticated than enjoying dialing tones.

In addition to everything else it does, your Android phone can play music. You can listen to songs you buy online, synchronize them from a computer, or find music on the Internet. Some phones can even listen to FM radio. So wherever you go with your phone, which should be everywhere, you also carry your music library.

'60s

Jazz

The Hits Just Keep On Comin'

Your Android phone is ready to entertain you with music whenever you want to hear it. Simply plug in the headphones, summon the music-playing app, and choose tunes to match your mood. It's truly blissful — well, until someone calls you and the phone ceases being a musical instrument and returns to being the ball-and-chain of the digital era.

To handle music-playing duties, your phone comes with a music-playing app. The stock Android app is Google's own Play Music app. That's the app covered in this chapter.

Browsing the music library

To view your music library, heed these directions:

1. **Start the Play Music app.**

2. **Tap the Side Menu icon to display the navigation drawer.**

   The Side Menu icon is found in the upper left corner of the screen. It's shown in the margin. If you see a left-pointing arrow instead, tap that arrow until the Side Menu icon appears.

3. **Choose My Library.**

The Play Music app is shown in Figure 14-1 with the My Library view selected. Your music is organized by categories, which appear as tabs atop the screen. Switch categories by tapping a tab, or swipe the screen left or right to browse your music library.

The categories make your music easier to find, because you don't always remember song, artist, or album names. The Genres category is for those times when you're in the mood for a certain type of music but don't know, or don't mind, who recorded it.

- ✔ Choose the Listen Now item from the navigation drawer to browse songs you frequently listen to or to discover tunes that the Play Music app guesses you'll like. The more you use the app, the more you'll appreciate the results shown in the Listen Now category.

- ✔ Songs and albums feature the Action Overflow icon, similar to the one shown in the margin. Use that icon to view a list of actions associated with the album or artist.

- ✔ Two types of album artwork are used by the Play Music app. For purchased music, or music recognized by the app, original album artwork appears. Otherwise, the app shows a generic album cover.

- ✔ When the Play Music app doesn't recognize an artist, it uses the title Unknown Artist. This usually happens with music you copy manually to your phone, but it can also apply to audio recordings you make.

Play Music notification   Categories   Action Overflow

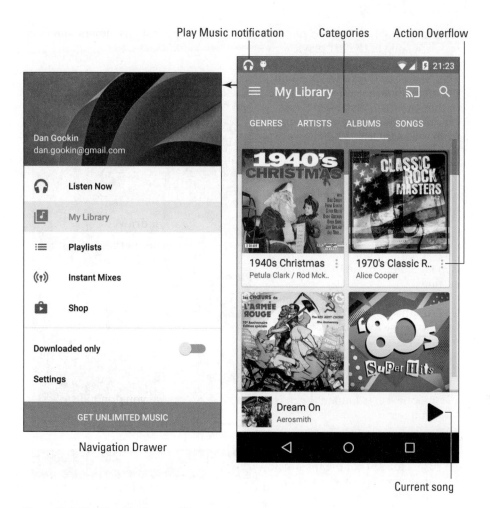

Navigation Drawer

Current song

Figure 14-1: The Play Music app; Albums category.

Playing a tune

Once you've found the proper tune to enhance your mood, play it! Tap on a song to play that song. Tap on an album to view songs in the album, or tap the album's large Play button, shown in the margin, to listen to the entire album.

While a song plays, controls appear at the bottom of the screen, as shown at the bottom right in Figure 14-1. Tap that card to view the song full-screen, as shown in Figure 14-2.

After the song has finished playing, the next song in the list plays.

The next song doesn't play if you have the Shuffle feature activated. In that case, the Play Music app randomly chooses another song from the same list. Who knows which one is next?

The next song might also not play when you have the Repeat option on: The three repeat settings are illustrated in Table 14-1, along with the Shuffle settings. To change settings, tap the Shuffle icon or the Repeat icon to cycle through the settings.

To stop the song from playing, tap the Pause icon, labeled in Figure 14-2.

✔ The volume is set by using the Volume key on the side of the phone: Up is louder, down is quieter.

✔ Music on your Android phone is played by streaming it from the cloud. That means music may won't play when an Internet connection is unavailable.

✔ You can keep music on your phone, playable with or without an Internet connection, by downloading it. Directions are offered in Chapter 16.

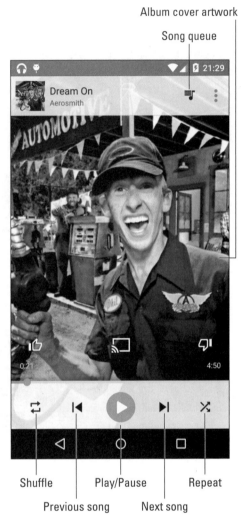

Album cover artwork

Song queue

Dream On
Aerosmith

Shuffle   Play/Pause   Repeat

Previous song   Next song

Figure 14-2: A song is playing.

✔ Use the Play Music app's Search icon to help locate tunes in your music library. You can search by artist name, song title, or album. After tapping the Search icon (refer to Figure 14-1), type all or part of the text you're searching for. Tap the Search key on the onscreen keyboard to start the search. Choose the song you want to hear from the list that's displayed.

✔ When a song is playing or paused, its album artwork might appear as the Lock screen wallpaper. Don't let the change alarm you.

Table 14-1	Shuffle and Repeat Icons	
Icon	Setting	Effect
	Shuffle is off.	Songs play one after the other.
	Shuffle is on.	Songs play in random order.
	Repeat is off.	Songs don't repeat.
	Repeat current song.	The same song plays over and over.
	Repeat all songs.	All songs in the list play over and over.

Controlling the music

The Play Music app keeps you audibly amused until the last song in the list plays, or forever when the Repeat All option is chosen. You're free to do anything else on the phone while music plays: open another app, read an eBook, or even lock the phone and let the music continue.

The only time music is interrupted is for an incoming call. After the call ends, you must restart the song: Visit the Play Music app and tap the Play icon.

 You can quickly access music controls by accessing the Play Music notification, shown in the margin. Play/Pause and Previous/Next icons are found on the notification, or you can tap the notification to return to the Play Music app.

Queuing up the next song

It's fun to randomly listen to your music library, plucking out tunes like a mad DJ. Oftentimes, however, you don't have patience to wait for the song to finish before choosing the next tune. The solution is to add songs to the queue. Here's how:

1. **Browse your music library for the next song (or album) you want to play.**

2. **Tap the song's Action Overflow.**

3. **Choose Add to Queue.**

   The Play Music app marks the song as next in line after the current song ends — and when any other queued songs end.

Songs are added to the queue in the order you tap them.

To review the queue, tap the Song Queue icon, shown in the margin as well as in Figure 14-2. The queue is shown in Figure 14-3.

Songs in the queue play in order, top down. To change the order, drag a song card up or down. To remove a song from the queue, swipe its card left or right.

If you like your queue, consider making a playlist of those same songs. See the section "Creating a playlist from a song queue," later in this chapter.

Being the life of the party

You need to do four things to make your Android phone the soul of your next shindig or soirée:

- ✔ Connect it to external speakers.
- ✔ Use the Shuffle setting.
- ✔ Set the Repeat option.
- ✔ Provide plenty of drinks and snacks.

The external speakers can be provided by anything from a custom media dock or a stereo to the sound system at the local sports stadium. You need an audio cable with a mini-headphone connector for the phone's headphone jack and an audio jack that matches the output device. Look for any store where the employees wear name tags.

After you connect your phone to the speakers, start the Play Music app. Choose the party playlist you've created, per the directions in the section "Organize Your Tunes," found elsewhere in this chapter.

Enjoy your party, and please drink responsibly.

Song Queue

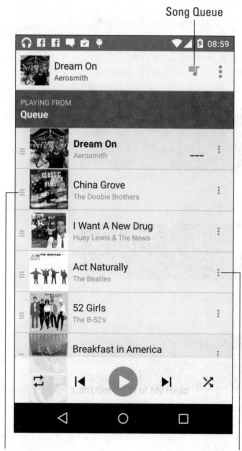

Drag to re-order songs   Song Action Overflow

Figure 14-3: The song queue.

"What's this song?"

You might consider getting a handy, music-oriented widget called Sound Search for Google Play. You can obtain this widget from the Google Play Store and then add it to the Home screen, as described in Chapter 19. From the Home screen, you can use the widget to identify music playing within earshot of your phone.

To use the widget, tap it on the Home screen. The widget immediately starts listening to your surroundings, as shown in the middle of the sidebar figure. After a few seconds, the song is recognized and displayed. You can choose to either buy the song at the Google Play Store or tap the Cancel icon and start over.

(continued)

(continued)

The Sound Search widget works best (exclusively, I would argue) with recorded music. Try as you might, you cannot sing into the thing and have it recognize a song. Humming doesn't work, either. I've tried playing the guitar and piano and — nope — that didn't work either. But for listening to ambient music, it's a good tool for discovering what you're listening to.

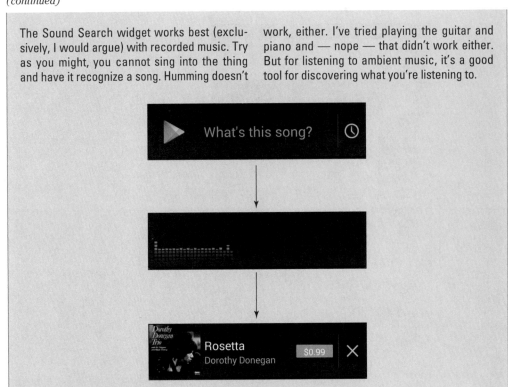

More Music for Your Phone

Your Android phone already has some music preinstalled, thanks to Google and the Play Music service. This random sampling of tunes may not be to your liking, and it's definitely not the final say in how much music you can have on your phone.

To add more music, you have two options:

- ✔ Buy lots of music from the Google Play Store, which is what Google wants you to do.
- ✔ Borrow music from your computer, which Google also wants you to do, just not as enthusiastically as the first option.

For information on buying music at the Play Store, see Chapter 15. To get music from your computer to the Google cloud and make it available to your phone, follow these steps:

1. **On your computer, locate the music you want to upload to your Google Play Music Library.**

   You can open a music jukebox program, such as Windows Media Player, or just have a folder window open that lists the songs you want to copy.

2. **Open the computer's web browser.**

3. **Visit** `https://music.google.com`.

   You see a copy of your Play Music library, including your playlists and any recently played songs. You can even listen to your music right there on the computer, but no: You have music to upload.

4. **Click the Upload Music button.**

5. **If you haven't yet done so, download the Music Manager.**

   Click the Download Music Manager button. Complete the installation process. You need to perform this task only once.

6. **Drag music into the web browser window, or click the Select from Your Computer button to add tunes.**

   It takes a while for Google to digest the songs, so be patient.

You can repeat these steps to upload tens of thousands of songs. The limit was once 25,000, but I believe Google increased that number recently.

The songs you upload are available to your Android phone, just like any other songs in the music library.

Organize Your Tunes

The Play Music app categorizes your music by album, artist, song, and so forth, but unless you have only one album and enjoy all the songs on it, that configuration probably won't do. To better organize your music, you can create *playlists*. That way, you can hear the music you want to hear, in the order you want, for whatever mood hits you.

Reviewing your playlists

To view any playlists that you've already created, or that have been preset on the phone, choose Playlists from the Play Music app's navigation drawer (refer to Figure 14-1). Playlists you've created are displayed on the screen, similar to what's shown in Figure 14-4.

To see which songs are in a playlist, tap the playlist's album icon. To play the songs in the playlist, tap the first song in the list.

Figure 14-4: Playlists in the Play Music app.

 A *playlist* is a helpful way to organize music when a song's information may not have been completely imported into the phone. For example, if you're like me, you probably have a lot of songs labeled Unknown. A quick way to remedy that situation is to organize those unknown songs on the playlist. The next section describes how it's done.

Creating a playlist

The Play Music app features three "auto" playlists, created automatically by the Play Music app:

Thumbs Up, which lists songs you've liked by tapping the thumbs-up icon

Last Added, which includes songs recently purchased or added to the phone

Free and Purchased, which includes just about everything

To create your own, custom playlist, follow these steps:

1. **Locate the song you want to add to a playlist.**

   You can also add an entire album of songs by choosing an album.

2. **Tap the song's Action Overflow icon.**

3. **Choose Add to Playlist.**

   Ensure that you're viewing a song or an album; otherwise, the Add to Playlist action isn't available.

4. **Choose New Playlist.**

   The New Playlist card appears.

5. **Fill in the New Playlist card.**

   Type the playlist's name and a description. You can set the playlist to be public by activating the Public master control. Such playlists can be shared with others on the Internet.

6. **Tap the Create Playlist button.**

   The playlist is created and stocked with the song or album you selected in Step 1.

See the next section for details on adding more songs to the playlist.

✔ You can have as many playlists as you like on the phone and stick as many songs as you like into them. Adding songs to a playlist doesn't affect the phone's storage capacity.

✔ To delete a playlist, tap the Action Overflow icon in the playlist's card (refer to Figure 14-1). Choose Delete and tap OK to confirm.

Adding songs to a playlist

To add new songs or albums to an existing playlist, follow these steps:

1. **Tap the Action Overflow icon on the song or album's card.**

2. **Choose Add to Playlist.**

3. **Select the playlist from the list on the Add to Playlist card.**

Repeat these steps to build up existing playlists.

✔ To remove a song from a playlist, open the playlist and tap the Action Overflow icon by a song's card. Choose Remove from Playlist.

✔ You can also swipe a song's card left or right to remove it from the playlist.

> ✔ Removing a song from a playlist doesn't delete the song from the music library; see the later section "Deleting music."
>
> ✔ Songs in a playlist can be rearranged: Drag the song's card up or down in the list.

Creating a playlist from a song queue

If you've created a song queue, and it's a memorable one, consider saving that queue as a playlist that you can listen to over and over. Obey these directions:

1. **Tap the Song Queue icon to view the song queue.**

   Refer to the earlier section "Queuing up the next song" for details on the song queue.

2. **Tap the Action Overflow icon and choose Save Queue.**

3. **Choose New Playlist.**

   Or you can add the songs to an existing playlist, as described in the preceding section.

4. **Fill in the New Playlist card.**

   Refer to the earlier section "Creating a new playlist."

5. **Tap the Create Playlist button.**

The songs in the current queue now dwell in their own playlist, always accessible.

Deleting music

To remove a song or an album from your Play Music library, tap its Action Overflow icon and choose Delete. Tap the OK button to remove the song. Bye-bye, music.

I don't recommend removing music. Most music on your phone is actually stored in the cloud, by Google's Play Music service. Therefore, removing the music doesn't affect the phone's storage. So unless you totally despise the song or artist, removing the music has no effect.

Some music can be stored locally by downloading it to the phone. That way, the music is always available. If the phone's storage capacity is a concern, you can unload the music from the phone. See Chapter 16 for details.

Music from the Stream

Although they're not broadcast radio stations, some sources on the Internet — *Internet radio* sites — play music. These Internet radio apps are available from the Google Play Store. Some free services that I can recommend are

- Pandora Radio
- Spotify
- TuneIn Radio

Pandora Radio and Spotify let you select music based on your mood and preferences. The more feedback you give the apps, the better the music selections.

The TuneIn Radio app gives you access to hundreds of Internet radio stations broadcasting around the world. They're organized by category, so you can find just about whatever you want. Many of the radio stations are also broadcast radio stations, so odds are good that you can find a local station or two, which you can listen to on your phone.

These apps, as well as other, similar apps, are available for free. Paid versions might also be found at the Google Play Store. The paid versions generally provide unlimited music with no advertising.

- Google offers an unlimited music listening service. You can sign up by tapping the Get Unlimited Music button, found at the bottom of the Play Music app's navigation drawer. The service is free for 30 days, and then a nominal fee, currently $9.99, is charged monthly.

- It's best to listen to Internet radio when your phone is connected to the Internet via a Wi-Fi connection. Streaming music can use a lot of your mobile data plan's monthly allotment.

- Be wary of music subscription services offered through the phone's manufacturer or cellular provider. I've subscribed to such services only to find them terminated for various reasons. To avoid that disappointment, stick with the services described in this section until you feel comfortable enough to buy into another service.

- Some phones may feature FM radio apps. These apps magically pull radio signals from the air and put them into your ear. The phone's radio hardware requires that a headset be plugged in for the app to work.

- Internet music of the type delivered by the apps mentioned in this section is referred to by the nerds as *streaming music*. That's because the music arrives on your phone as a continuous download from the source. The music is played as it comes in and isn't stored long-term.

Apps Various and Sundry

*W*hen cell phones started to become sophisticated, manufacturers began installing a few basic programs, or apps. Those early apps were pitiful. In fact, I remember my first cell phone that came with a smattering of boring, silly little apps. Still, it's kind of traditional for a phone to come with an alarm clock, a calculator, a calendar, and perhaps a few games. Your Android phone is no different.

The Alarm Clock

Your Android phone keeps constant and accurate track of the time, which is displayed at the top of the Home screen and also the Lock screen. That's handy, but it just isn't enough, so the phone ships with an app that tells the time and also may double as an alarm clock.

The app may be called Clock or Alarm. If it's called Alarm, then it's probably nothing more than a basic alarm clock. The Clock app, on the other hand, is more of a chronometric app, featuring a timer, stop watch, alarm, and world clock functions. Of these activities, setting an alarm is pretty useful: In that mode, your phone becomes your nightstand companion — and potentially your early morning nemesis.

To set an alarm in the Clock app, follow these generic steps:

1. **Tap the Alarm icon or tab atop the Clock app's screen.**

   The four tabs in the standard Android Clock app are Alarms, World Clock, Timer, and Stopwatch, although not specifically in that order.

2. **Tap the Add icon or ADD button.**

   A card appears, which you use to set the alarm time, days, name, and so on.

3. **Fill in details about the alarm.**

   Set the alarm's time. Determine whether it repeats daily or only on certain days. Choose a ringtone. Weigh over any other settings as shown on the card. The alarm name appears when the alarm triggers.

4. **Set the alarm.**

   Alarms must be set to activate.

When the alarm triggers, its name appears on the Lock screen. Tap the Dismiss icon and actually get out of bed. Or you can tap the Snooze icon to be annoyed again after a few minutes.

- ✔ When an alarm is set, an Alarm status icon appears atop the screen, similar to what appears in the margin. The icon is your clue that an alarm is set and ready to trigger.

- ✔ Un-setting an alarm doesn't delete the alarm. To remove an alarm, tap the alarm and then tap the Delete (trash) icon. In some incarnations of the Clock/Alarm app, long-press an alarm and choose the Delete action.

- ✔ The alarm doesn't work when you turn off the phone. However, the alarm does trigger when the phone is locked.

- ✔ For a larger time display, you can add the Clock widget to the Home screen. Refer to Chapter 19 for more information about widgets on the Home screen.

- ✔ So tell me: Do alarms go *off*, or do they go *on*?

The Calculator

The Calculator is perhaps the oldest of all traditional cell phone apps. It's probably also the least confusing and frustrating app to use.

The stock Android calculator app appears in Figure 15-1. The version you see on your phone may look different, although the basic operation remains the same. Also, consider changing the phone's orientation to see more or fewer buttons; the image in Figure 15-1 uses horizontal orientation, which shows the calculator's more frightening buttons.

Typical calculator buttons   Scary calculator buttons

Figure 15-1: The Calculator.

Type your equations by tapping the various buttons on the screen. Use the parenthesis to determine which part of a long equation gets calculated first. Use the DEL, CLR, or C button to clear input.

✔ Long-press the calculator's text (or results) to cut or copy the results. This trick may not work in every Calculator app.

✔ When a Clear All button isn't available, long press the DEL, CLR, or C button.

The Calendar

Once upon a time, people toted around a bulky notebook thing called a datebook. It assisted primitive humans with keeping a schedule, reviewing appointments, and knowing where they need to be and when. Such archaic technology is no longer necessary because your Android phone comes with a Calendar app.

✔ The Calendar app works with your Google account to keep track of your schedule and appointments. You can visit Google Calendar on the web at

```
http://calendar.google.com
```

✔ Before you throw away your datebook, copy into the Calendar app some future appointments and info, such as birthdays and anniversaries.

✔ The Calendar app may be called the S Planner app on some Samsung phones. It's the same thing.

Browsing dates and appointments

To check your schedule and browse upcoming events, open the Calendar app. You'll see upcoming dates shown in one of several views; Figure 15-2 shows the Calendar app's Month, Week, and Day views. Not shown are 3 Day view, Agenda view, or Schedule view, which simply lists upcoming events.

To change views, tap the Side Menu icon and choose a view type from the navigation drawer.

Swipe the screen left or right to browse events. If you need to return to today's date, tap the Go to Today icon, illustrated in Figure 15-2.

✔ Not every version of the Calendar app displays your schedule exactly as illustrated in Figure 15-2. Some versions may lack Week view or 3 Day view, for example. Others may sport an action bar from which the current calendar view is chosen.

✔ When using the 3 Day, Day, Week, or Schedule views, tap the month name to see a quick overview of the current month's appointments.

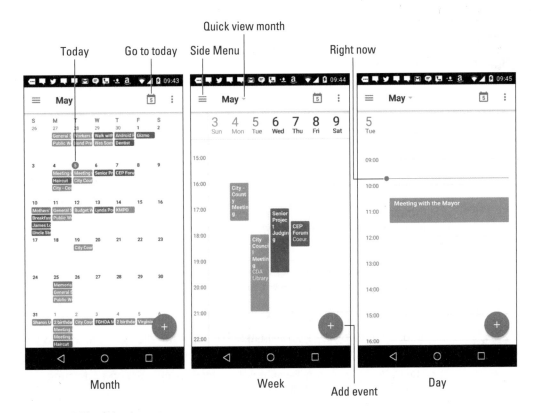

Figure 15-2: The Calendar app.

✔ I check Week view at the start of the week to remind me of what's coming up.

✔ The current day is highlighted in Month and Week views. A horizontal bar marks the current time (refer to Day view in Figure 15-2).

✔ Different colors flag your events. The colors represent a calendar category to which the events are assigned. See the later section "Creating an event," for information on calendars.

Reviewing appointments

To view more detail about an event, tap it. When you're using Month view, first tap the event's date and then the event. The event's details appear onscreen, similar to the event shown in Figure 15-3.

Which details you see depends on how much information was recorded when the event was created. Some events have only a minimum of information; others may have details, such as a location for the event. When the event's location is listed, you can tap that location, and the Maps app pops up to show you where the event is being held.

Tap the Back navigation icon to dismiss the event's details.

✔ Birthdays and a few other events on the calendar may be pulled from the phone's address book or from certain social networking apps. That probably explains why some events are listed twice; they're pulled in from two sources.

✔ The best way to review upcoming appointments is to choose Agenda view or Schedule view.

✔ The Calendar widget also provides a great way to see upcoming events from directly on the Home screen. See Chapter 19 for information on widgets.

✔ Google Now also lists any immediate appointments or events. See the later section "Google Now."

Creating an event

The key to making the calendar work is to add events: appointments, things to do, meetings, or full-day events such as birthdays or root canal work. To create an event, follow these steps in the Calendar app:

1. Select the day for the event.

Or, if you like, you can switch to Day view, where you can tap the starting time for the new event.

Edit event

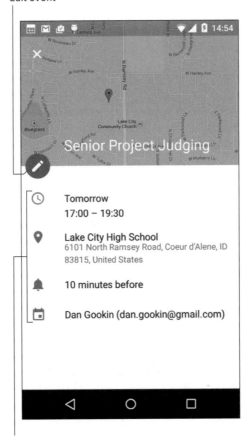

Event details

Figure 15-3: Event details.

2. Tap the New Event icon.

Upon success, you see a New Event card or an Add Event card. Your job is to fill in the blanks to create the event.

3. Set information for the event.

The more information you supply, the more detailed the event, and the more you can do with it on your phone. Here are some of the many items you can set when creating an event:

- *Title:* The name of the event, person you're going to meet, or place you're headed.

- *Calendar Category:* Choose a specific calendar to help organize and color-code your events.

- *Time/Duration:* If you followed Step 1 in this section, you don't have to set a starting time. Otherwise, specify the event's date, starting time and ending time. You can also set an all-day event such as a birthday or the cable installer's visit although they claimed he'd show up between 8:00 and 11:00.

- *Location:* Type the location just as though you're searching for a location in the Maps app.

- *Repeat:* Configure the event to happen regularly on a monthly or weekly schedule. If you don't see the item, tap the More Options button.

- *Notification/Reminder:* Set an email, text message, or Calendar notification to signal an upcoming event.

4. **Tap the Save or Done button to create the event.**

The new event appears on the calendar, reminding you that you need to do something on such-and-such a day.

When an event's date-and-time arrives, an event reminder notification appears, similar to the one shown in the margin. You might also receive a Gmail notification or text message, depending on how you chose to be reminded when the event was created.

✔ To change, update, or edit an event, tap the event to bring up its card as shown earlier in Figure 15-3. Tap the Edit icon to make modifications.

✔ For events that repeat twice a week or twice a month, create *two* repeating events. For example, when you have meetings on the first and third Mondays, you should create two separate events: one for the first Monday and another for the third. Then have each event repeat monthly.

✔ To remove an event, first edit it by tapping the Edit icon on the event's card. Tap the Delete button or the Delete (Trash) icon. Tap OK to confirm. For a repeating event, choose whether to delete only the current event or all future events.

✔ Setting an event's time zone is necessary only when the event takes place in another time zone or spans time zones, such as an airline flight. In that case, the Calendar app automatically adjusts the starting and stopping times for events, depending on where you are.

✔ If you forget to set the time zone and you end up hopping around the world, your events are set according to the time zone in which they were created, not the local time.

✔ Calendar categories are handy because they let you organize and color-code your events. They can be confusing because Google calls them "calendars." I think of them more as categories: I have different calendars (categories) for my personal and work schedules, government duties, clubs, and so on.

✔ Avoid using the Phone and Device categories for your events. Events in those categories appear on your phone but aren't shared with your Google account.

The eBook Reader

To sate your electronic book-reading desires, your Android phone comes with Google's eBook reader app, Play Books. It offers you access to your eBook library, plus the multitudinous tomes available at the Google Play Store

Open the Play Books app to behold any eBooks you've recently read, which are displayed on the Read Now screen. To browse your entire eBook library, tap the Side Menu icon and choose My Library from the navigation drawer.

The library lists any titles you've obtained for your Google Books account, similar to what's shown in Figure 15-4.

Scroll through the library by swiping the screen.

Tap a book's cover (or card) to open it. If you've previously opened the book, you see the page you last read. Otherwise, you see the book's first page.

Figure 15-5 illustrates the basic book-reading operation in the Play Books app. You turn pages by swiping left or right, but probably mostly left. You can also turn pages by tapping either the far left or right side of the screen.

The Play Books app works in both vertical and horizontal orientations. You can lock the screen orientation for the app to either direction: Choose the Settings item from the navigation drawer (refer to Figure 15-4), and then choose Auto-Rotate Screen. Select a Lock item to prevent the app from rotating the screen while you read.

Book's Action Overflow

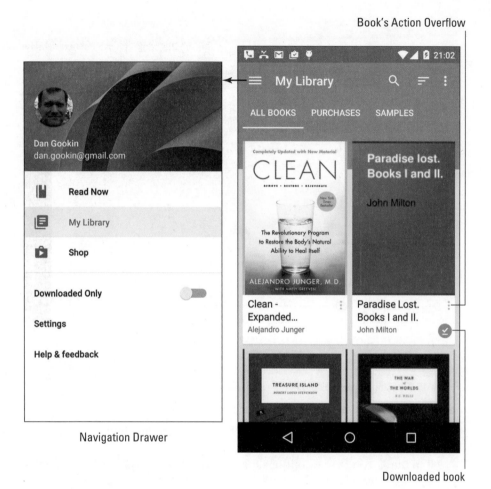

Navigation Drawer

Downloaded book

Figure 15-4: The Play Books library.

- ✔ If you don't see a particular book in your library, tap the Action Overflow and choose Refresh.

- ✔ Books in your Play Books library are stored on the Internet and available to read only when an Internet connection is active. It's possible to keep a book on your phone by downloading it to the device. Refer to Chapter 16 for details on downloading books.

- ✔ To remove a book from the library, tap the Action Overflow on the book's cover and choose Delete from Library.

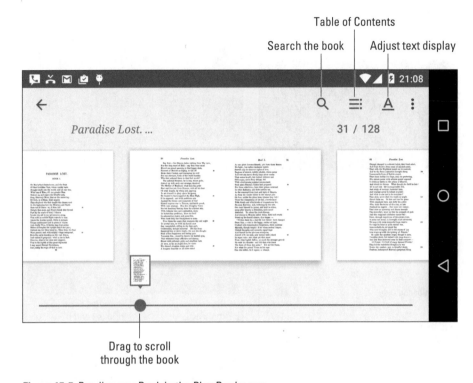

Table of Contents

Search the book · Adjust text display

Drag to scroll
through the book

Figure 15-5: Reading an eBook in the Play Books app.

✔ If the onscreen controls (refer to Figure 15-5) disappear, tap the screen to see them again.

✔ Tap the A icon to view options for adjusting the text on the screen and the brightness level.

✔ Unlike dead-tree books, eBooks lack an index. That's because text on digital pages can change based on the book's presentation. Therefore, use the Search icon (refer to Figure 15-5) to find text.

✔ Copies of all your Google Books are available on all your Android devices and on the `http://books.google.com` website.

✔ Refer to Chapter 16 for information on obtaining books at the Google Play Store, as well as how to download books to your phone.

✔ If you have a Kindle device, you can obtain the Amazon Kindle app for your phone. Use the app to access books you've purchased for the Kindle, or just as a supplement to Google Books.

The Game Machine

For all its seriousness and technology, one of the best uses of a smartphone is to play games. I'm not talking about silly arcade games (though I admit that they're fun). No, I'm talking about some serious portable gaming.

To whet your appetite, your Android phone may have come with a small taste of what the device can do in regard to gaming; look for preinstalled game apps in the apps drawer. If you don't find any, choose from among the hordes available at the Google Play Store, covered in Chapter 16.

Game apps use phone features such as the touchscreen or the accelerometer to control the action. It takes some getting used to, especially if you regularly play using a game console or PC, but it can be fun — and addicting.

Google Now

Don't worry about the phone controlling too much of your life: It harbors no insidious intelligence, and the Robot Uprising is still years away. Until then, you can use your phone's listening capabilities to enjoy the feature called Google Now. It's not quite like having your own personal Jeeves, but it's on its way.

Finding the Google Now feature can be fun! On some phones, you drag your finger on the touchscreen from bottom to top to summon it. Other phones may feature the Google Now screen as the far left Home screen page. Or you can summon Google Now by tapping the Google Search widget on the Home screen.

Figure 15-6 illustrates a typical Google Now screen. Cards appear below the search text box. The variety and number of cards depend on how often you use Google Now. The more it learns about you, the more cards appear.

You can use Google Now to search the Internet, just as you would use Google's main web page. More interesting than that, you can ask Google Now questions; see the nearby sidebar "Barking orders to Google Now."

Google Now might also dwell in the apps drawer. It might be named Google Now or just Google. On some phones, you may have to tap a Get Google Now link to obtain the app.

Search for something   Ask a question

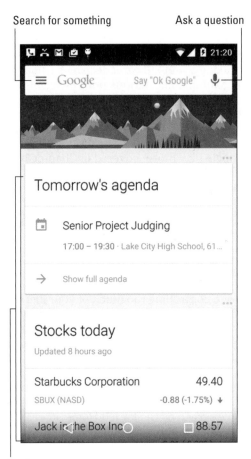

Cards

Figure 15-6: Google Now is ready for business. Or play.

Barking orders to Google Now

One way to have a lot of fun is to use the Google Now app verbally. Just say "OK Google." Say it out loud. Anytime you see the Google Now app, it's listening to you. Or when the app is being stubborn, tap the Microphone icon.

You can speak simple search terms, such as "Find pictures of Megan Fox." Or you can give more complex orders, among them:

- Will it rain tomorrow?
- What time is it in Frankfurt, Germany?
- How many euros equal $25?
- What is 103 divided by 6?
- How can I get to Disneyland?

- ✔ Where is the nearest Sri Lankan restaurant?

- ✔ What's the score of the Lakers–Celtics game?

- ✔ What is the answer to life, the universe, and everything?

When asked such questions, Google Now responds with a card as well as a verbal reply. When a verbal reply isn't available, Google search results are displayed.

The Video Player

It's not possible to watch "real" TV on your phone, but a few apps come close. The YouTube app is handy for watching random, meaningless drivel, which I suppose makes it a lot like TV. Google offers the Play Movies & TV app, which lets you buy and rent real movies and TV shows from the Play Store. And when you tire of those apps, you can use the Camera app with the front-facing camera to pretend that you're the star of your own reality TV show.

Watching YouTube

YouTube is the Internet phenomenon that proves that real life is indeed too boring and random for television. Or is that the other way around? Regardless, you can view the latest videos on YouTube — or contribute your own — by using the YouTube app on your Android phone.

Search for videos in the YouTube app by tapping the Search icon. Type the video name, a topic, or any search terms to locate videos. Zillions of videos are available.

The YouTube app displays suggestions for any channels to which you're subscribed, which allows you to follow favorite topics or YouTube content providers.

- ✔ Use the YouTube app to view YouTube videos, rather than use the phone's web browser app to visit the YouTube website.

- ✔ Orient the phone horizontally to view the video in a larger size.

- ✔ Because you have a Google account, you also have a YouTube account. I recommend that you log in to your YouTube account when using the YouTube app: Tap the Action Overflow and choose Sign In. You see your account information, your videos, and any video subscriptions.

- ✔ Not all YouTube videos are available for viewing on mobile devices.

Buying and renting movies

You can use the Play Movies & TV app to watch videos you've rented or purchased at the Google Play Store. Open the app and choose the video from the main screen. Items you've purchased show up in the app's library.

The actual renting or purchasing is done at the Google Play Store. Check there often for freebies and discounts. More details for renting and purchasing movies and shows is found in Chapter 16.

✔ Movies and shows rented at the Play Store are available for viewing for up to 30 days after you pay the rental fee. After you start the movie, you can pause and watch it again and again during a 24-hour period.

✔ Not every film or TV show is available for purchase. Some are rentals only.

✔ Also see Chapter 18 for information on screencasting video from your phone to a large-screen monitor or HDTV.

Shop at the Play Store

In This Chapter

▶ Shopping at the Play Store

▶ Getting an app, music, eBook, and so on

▶ Understanding the purchasing process

▶ Building a wish list

▶ Sending an app suggestion to a friend

▶ Keeping media on the phone

*Y*our Android phone may have come packed with diverse and interesting apps, or it may have only a paltry selection of tepid sample apps. Regardless of the assortment, you're in no way limited to the original collection. That's because hundreds of thousands of different apps are available for your phone: productivity apps, references, tools, games, and more. They're all found at the central shopping place for all Android phones: the Google Play Store.

Welcome to the Play Store

The Google Play Store is one of the things that makes owning an Android phone rewarding. You can visit the Play Store not only to obtain more apps for your phone but also to go shopping for music, books, and videos (films and TV shows). Someday, Google might even sell robots at the Play Store. Until then, the selection seems adequate and satisfying.

...ıoAddict Inc
★★★★★

Lookout Security & Antivirus
Lookout Mobile Security
★★★★★

 Google Hindi Input
Google Inc. ❖
★★★★☆

Backgrounds HD Wallpapers
Stylem Media
★★★

▶ Officially, the store is the Google Play Store. It may also be referenced as Google Play. The app is named Play Store.

▶ The Google Play Store was once known as Android Market, and you may still see it referred to as the Market.

TIP

✔ You obtain more goodies from Play Store over an Internet connection. Therefore:

✔ I highly recommend that you connect your phone to a Wi-Fi network if you plan to obtain apps, books, or movies at the Play Store. Wi-Fi not only gives you speed, but it also helps avoid data surcharges. See Chapter 17 for details on connecting your phone to a Wi-Fi network.

✔ The Play Store app is frequently updated, so its look may change from what you see in this chapter. Refer to my website for updated info and tips: www.wambooli.com/help/android.

Browsing the Play Store

You access the Google Play Store by opening the Play Store app, found in the Apps drawer. You may also find a launcher on the Home screen.

After opening the Play Store app, you see the main screen, similar to the one shown in Figure 16-1. If not, tap the Side Menu icon (shown in the margin) to display the navigation drawer. Choose Store Home to see the app's main screen.

To browse, tap a category atop the screen. Categories available include apps, games, movies and TV, music, books, and magazines. Suggestions are offered below the main categories, as shown in Figure 16-1. Swipe the suggestions up and down to peruse the lot.

After you choose a category, the next screen lists popular and featured items, plus additional categories you can browse by swiping the screen from right to left. Category titles appear toward the top of the screen. These include topics such as Top Free, New Movie Releases, Top Songs, and Deals.

When you have an idea of what you want, such as an album name or an app's function, searching works fastest: Tap the Search icon at the top of the Play Store screen, shown in Figure 16-2. At the main Play Store screen, tap the Google Play box to search (see Figure 16-1). As you type, the app lists suggestions.

Tap an item to see a more detailed description, screen shots, a video preview, comments, and links to similar items, as shown in Figure 16-2.

✔ The first time you enter the Google Play Store, or after the Play Store app is updated, you have to accept the terms of service. To do so, tap the Accept button.

✔ You can be assured that all apps that appear in the Play Store can be used with your phone. There's no way to obtain something that's incompatible.

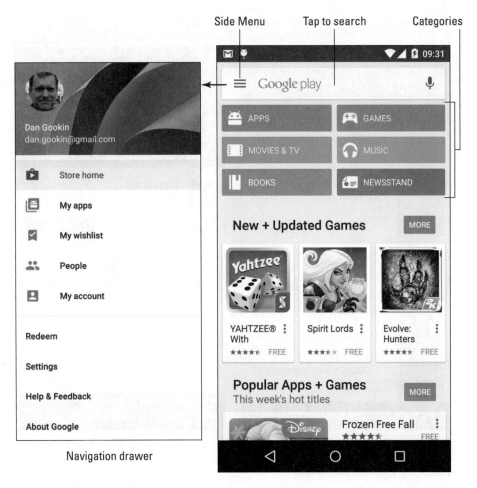

Figure 16-1: The Google Play Store.

✔ Pay attention to an app's ratings. Ratings are added by people who use the apps — people like you and me. Having more stars is better. You can see additional information, including individual user reviews, by choosing the app.

✔ Another good indicator of an app's success is how many times it's been downloaded. Some apps have been downloaded tens of millions of times. That's a good sign.

✔ In Figure 16-2, the app's description (on the right) shows an Install button. Other buttons that may appear on an app's description screen include Open, Update, Refund, and Uninstall. The Open button opens an app that's already installed on your phone. Refund is available just after you purchase something. See Chapter 19 for information on using the Update and Uninstall buttons.

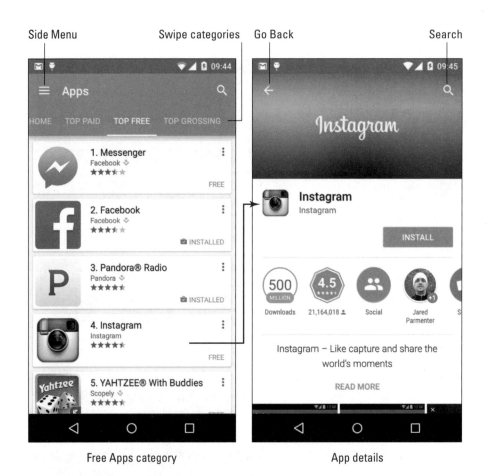

Free Apps category   App details

Figure 16-2: Hunting down an app.

Obtaining an item

After you locate an app, some music, or a book at the Play Store, the next step is to download it into your phone. The item is installed automatically, expanding what your phone can do or building your media library.

Good news: Most apps are free. Classic books are available at no cost. And occasionally, Google offers movies and music *gratis*. Even the items you pay for don't cost that much. In fact, it seems odd to sit and stew over whether paying 99 cents for a game is "worth it."

I recommend that you download a free app or book first, to familiarize yourself with the process. Then try your hand at downloading a paid item.

Free or not, the process of obtaining something from the play store works pretty much the same. Follow these steps:

1. **If possible, activate the Wi-Fi connection to avoid incurring data overages.**

   The phone uses Wi-Fi when it's available, preferring it over the mobile data network. See Chapter 17 for information on connecting your phone to a Wi-Fi network.

2. **Open the Play Store app.**

3. **Find the item you want and open its description.**

   All items in the Play Store feature a description screen. It looks similar to the app description screen shown on the right in Figure 16-2.

   The difference between a free item and one that costs money is found on the button. For a free items, the button says *Install*. For paid items, the button shows the price.

4. **Tap the Install button to get a free app or the Add to Library button for a free book or another item; for a paid app, tap the Buy (Price) button.**

   You may also see a Free Trial button for some items. Tap that button to obtain a brief sample of media (books, music, movies).

5. **For an app, tap the Accept button.**

   The access card describes which phone features the app uses. The list isn't a warning, and it doesn't mean anything bad. Still, see the nearby sidebar "Avoiding Android viruses."

6. **For a paid item, tap the Buy button.**

   You may have to confirm the payment method. See the next section for details.

   Media items are instantly available. Apps are downloaded to your phone, which may take some time. You see the Downloading notification appear atop the screen, similar to the one shown in the margin.

   You're free to do other things on your phone while an app is downloaded and installed.

7. **Tap the Open, Play, Listen, Read, or similar button to run the app, watch a video, listen to music, or read a book, respectively.**

Music, eBooks, movies, and videos are available for instant enjoyment, but not necessarily downloaded to your phone. See the later section "Keeping stuff on the device."

✔ To quickly access something you've just acquired from the Play Store, look for the Successfully Installed notification, shown in the margin. The notification features the item's name with the text *Successfully Installed* beneath it.

✔ For some apps, you may have to accept a license agreement. If so, tap the I Agree button. Additional app setup may involve setting your location, signing in to an account, or creating a profile, for example.

✔ Eventually, you receive a Gmail message from the Google Play Store, confirming your purchase, paid or free. The message contains a link you can select to review the app refund policy, in case you change your mind about the purchase.

✔ Be quick on that refund: For a purchased app, you have only two hours to get your money back. You know when the time limit is up because the Refund button on the app's description screen changes to Uninstall.

✔ The Play Store doesn't currently offer refunds on purchased media, which includes music, books, and movies.

✔ Keep an eye out for special offers at the Play Store. A special offer is a great way to pick up some free songs and albums.

Purchasing something at the Play Store

To purchase an app, music, eBook, or anything else at the Play Store, you tap the Buy button. A card appears, listing available payment methods. These include any credit cards you've used previously, plus any redeemed Google credit.

Figure 16-3 shows a typical Buy card. The album costs $5.99. The chosen payment method is a Visa card ending in 1869. That payment method is used after the Buy button is tapped.

Avoiding Android viruses

How can you tell which apps are legitimate and which might be viruses or evil apps that do odd things to your Android phone? Well, you can't. In fact, most people can't, because evil apps don't advertise themselves as such.

The key to knowing whether an app is malicious is to look at its access card: Open the app's description in the Play Store, and see which permissions it's granted. For example, if a simple grocery-list app uses the phone's microphone and the app doesn't need to use the microphone, it's suspect.

In the history of the Android operating system, only a handful of malicious apps have been

distributed, and most of them were found on devices used in Asia. Google routinely removes malicious apps from the Play Store. Google is also capable of remotely wiping such apps from all Android devices. So you're pretty safe.

Generally speaking, avoid "hacker" apps, porn apps, and apps that use social engineering to make you do things on your phone that you wouldn't otherwise do, such as visit an unknown website to see racy pictures of politicians or celebrities.

Current payment method

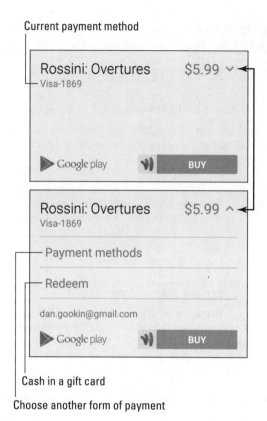

Cash in a gift card

Choose another form of payment

Figure 16-3: The Play Store's Buy card.

To select another payment method, follow these steps when the Buy card is presented:

1. **Tap the chevron by the price.**

   The chevron is shown next to the price *$5.99* at the top of Figure 16-3.

2. **Choose Payment Methods.**

   Of, if you're fortunate enough to have a Google Play gift card, tap Redeem to cash in.

3. **Choose a credit or debit card, PayPal, or a gift card.**

   The credit or debit cards listed are those you've used before. Don't worry: Your information is safe.

   If you don't yet have a credit or debit card registered, choose the Add Credit or Debit Card option and complete the onscreen steps to set up your payment.

4. **Type your Google password.**

This step provides for security.

I strongly recommend that you *do not* choose the option Never Ask Me Again.

5. **If prompted, tap the Every Time option.**

By choosing this option, you ensure that your password is required for every purchase you make at the Play Store. I recommend that you make this setting if it's presented.

At this, point, the item you purchased is made available. Apps are downloaded and installed; music, eBooks, movies, and videos are available but not necessarily downloaded. See the later section "Keeping stuff on the device."

Renting or purchasing videos

When it comes to movies and TV shows available at the Google Play Store, you have two options: rent or purchase.

When you choose to rent a video, the rental is available to view for the next 30 days. After you start watching, however, you have only 24 hours to finish — you can also watch the video over and over again during that time span.

Purchasing a video is more expensive than renting it, but you can view the movie or TV show at any time, on any Android device. You can also download the movie so that you can watch it even when an Internet connection isn't available.

Never buy anything twice

Any items you've already purchased in the Google Play Store are available for download on your current Android phone. Books, music, videos, and other media are accessed via the appropriate app, such as Play Books and Play Music. Apps you've purchased must be downloaded: Locate the app in the Play Store and tap the Install button.

To review any already purchased apps, display the Play Store app's navigation drawer (refer to the left side of Figure 16-1). Choose My Apps. Choose the All tab from the top of the screen. You see all the apps you've ever obtained at the Google Play Store, including apps you've previously paid for. Those apps are flagged with the text *Purchased*. Choose that item to reinstall the paid app.

One choice you must make when buying a movie is whether to purchase the SD or HD version. The SD version is cheaper, and the HD version plays at high definition only on certain output devices. Obviously, when watching on your phone only, the SD option is preferred.

Also see Chapter 15 for information on watching movies and TV shows.

Play Store Tricks

Like most people, you probably don't want to become a Play Store expert. You just want to get the app you want and get on with your life. When you're ready to get more from the Play Store, peruse the items in this section.

Using the wish list

 While you dither over getting a purchase at the Play Store, consider adding the item to your wish list: Tap the Wish List icon when viewing the app. The Wish List icon is shown in the margin, although its color changes depending on which category you're viewing in the Play Store app.

To review your wish list, tap the Side Menu icon in the Play Store app (refer to Figure 16-1). Choose the My Wishlist item from the navigation drawer. You'll see all the items you've flagged. When you're ready to buy, choose one and buy it!

Sharing an item from the Play Store

Sometimes you love your Play Store purchase so much that you just can't contain your glee. When that happens, consider sharing the item. Obey these steps:

1. **Open the Play Store app.**

2. **Browse or search for the app, music, book, or other item you want to share.**

3. **When you find the item, tap it to view its description screen.**

4. **Tap the Share icon.**

    You may have to swipe down the screen to locate the Share icon, shown in the margin. After tapping the Share icon, you see a menu listing various apps.

5. **Choose an app.**

   For example, choose Gmail to send a Play Store link in an email message.

6. **Use the chosen app to share the link.**

   What happens next depends on which sharing method you've chosen.

The result of following these steps is that your friend receives a link. That person can tap the link on his mobile Android device and be whisked instantly to the Google Play Store, where the item can be obtained.

Keeping stuff on the device

Books, music, movies, and TV shows that you obtain from the Play Store are not copied to your phone. Instead, they're stored on the Internet. When you access the media, it's streamed into your device as needed. This setup works well, and it keeps your phone from running out of storage space, but it works only when an Internet connection is available.

When you plan on being away from an Internet connection, such as when you are flying cross-country and are too cheap to pay for inflight Wi-Fi, you can download Play Store music, eBook, and movie purchases and save them on your phone.

To see which media is on your phone and which isn't, open the Play Books, Play Music, or Play Movies & TV app. Follow these steps, which work identically in each app:

1. **Tap the Side Menu icon.**

2. **In the navigation drawer, locate the Downloaded Only item and slide its master control to the On position.**

   Just tap the gizmo, and it toggles between the On and Off settings.

3. **Choose the My Library item from the navigation drawer.**

   You see only those items on your phone. The rest of your library, you can assume, is held on the Internet.

To see your entire library again, repeat these steps but in Step 2 slide the master control to the Off position.

 Items downloaded to your phone feature the On Device icon, similar to the one shown in the margin. The icon's color differs between music, eBooks, and movies.

 To keep an item on your phone, look for the Download icon, shown in the margin. Tap that icon, and the item is fetched from the Internet and stored on your device.

Keeping movies and lots of music on your Android phone consumes a lot of storage space. That's okay for short trips and such, but for the long term, consider purging some of your downloaded media: Tap the On Device icon. Tap the Remove button to confirm.

Don't worry about removing downloaded media. You can always access items you've purchased (or obtained for free) when an Internet connection is active. And you can download items over and over without having to pay again.

Part IV
Nuts and Bolts

In this part . . .

- ✔ Seek out and find Wi-Fi networks, connect, and use the Internet.
- ✔ Exchange files between your phone and other devices.
- ✔ Work with apps and widgets on the Home screen.
- ✔ Customize and configure your phone.
- ✔ Keep your phone private and secure.
- ✔ Take your phone abroad.
- ✔ Maintain and troubleshoot your phone.

No Wires, Ever!

portable implies that something can be moved, but not how far or how easily. My first TV was a "portable," which meant that it had handholds on the sides; the TV weighed about 25 pounds. *Cordless* implies a degree of freedom beyond portable, but still requires a charging station. Cordless devices typically have a range beyond which they fail to communicate. *Wireless* is at the top of the mobility heap. It implies freedom from wires but is also lightweight and tiny.

Your Android phone is truly wireless. Forget for a second that you need a wire to charge it, and even that chore is optional. After the battery is charged, you can take your phone anywhere and use it wire-free. You can access the mobile network, a Wi-Fi network, and even wireless gizmos and peripherals. This chapter describes the degrees to which you can keep your phone wireless.

mperial Wambooli
Connected

CDA COTTAGE
Secured with WPA2 (WPS available)

LuckyDog
Secured with WPA/WPA2 (WPS available)

yEagle
ith WPA/WPA2 (WPS

Wireless Networking Wizardry

You know that wireless networking has hit the big-time when you see people asking Santa Claus for a Wi-Fi router at Christmas. Such a thing would have been unheard of years ago because routers were used primarily for woodworking back then. Today, wireless networking is what keeps gizmos like your Android phone connected to the Internet.

Using the mobile data network

You pay your cellular provider a handsome fee every month. The fee comes in two chunks. One chunk (the less expensive of the two) is the telephone service. The second chunk is the data service, which is how your Android phone gets on the Internet. This system is the *mobile data network*.

The type of mobile data network the phone uses is indicated by a status icon atop the touchscreen, right next to the signal-strength bars. Here's the gamut:

- ✔ **4G LTE:** This is the fourth generation of wide-area data networks and the fastest.

- ✔ **3G:** It's the third-generation mobile data network.

- ✔ **1X:** Several types of the original, slower cellular data signals are still available. They all fall under the 1X banner. It's slow.

Your phone always uses the best network available. So when a 4G LTE network is within reach, that network is used for Internet communications. Otherwise, the 3G network is chosen, and then 1X as a form of last-ditch desperation.

- ✔ When a data network isn't available, no icon appears on the status bar. In fact, it's entirely possible for the phone to have no data signal but still be able to make phone calls.

- ✔ See Chapter 24 for information on how to monitor mobile data usage and avoid surcharges.

- ✔ When both a mobile data network and Wi-Fi are available, your phone uses Wi-Fi for all Internet access. Therefore, I recommend connecting to and using a Wi-Fi network wherever possible.

Understanding Wi-Fi

The mobile data connection is nice and is available pretty much all over, but it costs you money. A better option, and one you should seek out when it's available, is *Wi-Fi,* or the same wireless networking standard used by computers for communicating with each other and the Internet.

Making Wi-Fi work on your Android phone requires two steps. First, you must activate the phone's Wi-Fi radio. The second step is connecting to a specific wireless network. The next two sections cover the details.

Wi-Fi stands for *wi*reless *fi*delity. It's brought to you by the numbers 802.11 and various letters of the alphabet too many to mention.

Activating Wi-Fi

Follow these steps to activate Wi-Fi on your Android phone:

1. **Open the Settings app.**

2. **Choose Wi-Fi.**

   On some Samsung phones, tap the Connections tab to locate the Wi-Fi item.

3. **Ensure that the Wi-Fi master control icon is on.**

To turn off Wi-Fi, repeat the steps in this section, but in Step 3 slide the master control to Off. Turning off Wi-Fi disconnects your phone from any wireless networks.

✔ You can quickly peruse your phone's Wi-Fi settings by choosing the Wi-Fi quick setting. See Chapter 3 for more information on Quick Settings.

✔ It's perfectly okay to keep the phone's Wi-Fi radio on all the time. It is not a major drain on the battery.

✔ Using Wi-Fi to connect to the Internet doesn't incur data usage charges.

Connecting to a Wi-Fi network

After activating the Wi-Fi radio on your Android phone, you can connect to an available wireless network. Any network you've previously connected to is automatically reconnected. Otherwise, you can connect to an available network by following these steps:

1. **Open the Settings app.**

2. **Choose Wi-Fi or Wireless & Networks.**

   On some Samsung phones, the Wi-Fi item is found on the Connections tab.

3. **Choose a wireless network from the list.**

   Available Wi-Fi networks appear on the screen, similar to what's shown in Figure 17-1. When no wireless networks are listed, you're sort of out of luck regarding wireless access from your current location.

4. **If prompted, type the network password.**

   Tap the Show Password box so that you can see what you're typing. That helps when the password is intolerably long or complex.

5. Tap the Connect button.

The network is connected immediately. If not, try the password again.

Wi-Fi Master Control

Password-protected network

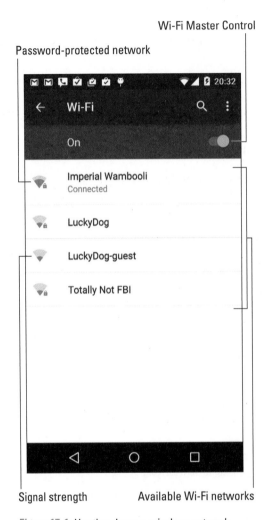

Signal strength   Available Wi-Fi networks

Figure 17-1: Hunting down a wireless network.

 When the phone is connected, you see the Wi-Fi status icon atop the touch-screen, looking similar to the icon shown in the margin. This icon indicates that the phone's Wi-Fi is on — connected and communicating with a Wi-Fi network.

- ✔ Some public networks are open to anyone, but you have to use the phone's web browser app to find a login web page. Simply browse to any page on the Internet, and the login web page shows up. Follow the directions to get network access.

- ✔ Not every wireless network has a password. They should! I don't avoid connecting to any public network that lacks a password, but I don't use that network for shopping, banking, or any other secure online activity.

- ✔ Unlike the mobile network, a Wi-Fi network's broadcast signal goes only so far. My advice is to use Wi-Fi when you plan to remain in one location for a while. If you wander too far away, your phone loses the signal and is disconnected.

Connecting to a hidden Wi-Fi network

Some wireless networks don't broadcast their names, which adds security but also makes accessing them more difficult. In these cases, follow these steps to connect to the hidden Wi-Fi network:

1. **Open the Settings app and choose Wi-Fi.**

2. **Tap the Action Overflow and choose Add Network.**

   The item might be titled Add Wi-Fi Network. Some phones may use an Add (plus sign) icon as the Add Network action.

3. **Type the network name into the Enter the SSID box.**

4. **Choose the security setting.**

5. **Type the password.**

   The password may be optional. See the preceding section or my advice on using password-less public networks.

6. **Tap the Save button or Connect button.**

Obtain the SSID, security, and password information from the girl with the pink hair and pierced lip who sold you coffee or from whoever is in charge of the wireless network at your location.

SSID stands for Service Set Identifier. It is not considered a valid acronym for use in Scrabble, despite its pitiful 5 points.

Connecting to a WPS router

Many Wi-Fi routers feature WPS, which stands for Wi-Fi Protected Setup. It's a network authorization system that's really simple and quite secure. If the wireless router features WPS, you can use it to quickly connect your phone to the network.

To make the WPS connection, follow these steps:

1. **Open the Settings app and choose Wi-Fi.**

2. **If you can't find the WPS option or icon on the screen, tap the Action Overflow icon and choose Advanced.**

   Two WPS options are available, depending on how the router works. Some WPS routers are push-button activated; others list a PIN.

3. **Choose WPS Push Button or WPS Pin Entry, depending on the router.**

4. **Push the router's WPS button or, on your phone, type the router's PIN.**

   The router's WPS button is labeled by using the letters *WPS* or the WPS icon, shown in the margin.

Connection with the router may take a few minutes, so be patient. The good news is that, like all Wi-Fi networks, after the initial connection is established, the phone automatically connects in the future.

Share the Connection

Your Android phone mobile data connection is for more than just your own phone. It's possible to share that Internet access in one of two ways. The first is to create a mobile *hotspot,* which allows any Wi-Fi–enabled gizmo to access the Internet via your phone. The second is a direct connection between your phone and another device, which called *tethering.*

Creating a mobile hotspot

To share your phone's mobile data signal with other Wi-Fi devices in the vicinity, heed these steps:

1. **Open the Settings app.**

2. **Turn off the phone's Wi-Fi radio.**

   There's no point in creating a Wi-Fi hotspot when one is already available.

3. **Plug the phone into a power source.**

   The mobile hotspot feature draws a lot of power.

4. **Tap the Back navigation icon to return to the main Settings app screen.**

5. **Choose the More item.**

   You'll find the More item in the Wireless & Networks section of the Settings app. On some Samsung phones, locate the More item on the Connections tab.

6. **Choose Tethering & Portable Hotspot.**

   This item might be titled Mobile Hotspot.

7. **Slide the Portable Wi-Fi Hotspot master control to the On position.**

   On some phones, the item may be titled Mobile Hotspot. It might also feature a check box instead of a master control.

   After the hotspot is active, you need to confirm some of the settings.

8. **Choose Set Up Wi-Fi Hotspot.**

   On some Samsung phones, tap the MORE button and choose Configure Mobile Hotspot.

   Tap the fields on the Set Up Wi-Fi Hotspot card to assign a name, security level, and password.

9. **Tap the Save button or OK button to set your changes.**

The phone's Wi-Fi hotspot can be accessed by any other mobile device or computer with Wi-Fi networking.

While the mobile hotspot is active, a Hotspot Active status icon appears, similar to the one shown in the margin.

To turn off the mobile hotspot, repeat the steps in this section, but slide the master control to the Off position in Step 7.

- Your phone's ability to create a mobile hotspot can be limited by the cellular provider. Some require an extra subscription (fee) before you can access this feature.

- Some phones may feature a Mobile Hotspot or 4G Hotspot app. If so, use it instead of following the steps in this section.

- The range for the mobile hotspot is about 30 feet. Items such as walls and tornadoes can interfere with the signal, rendering it much shorter.

- The default Wi-Fi hotspot password is your phone number.

- When you use the mobile hotspot, data usage fees apply. When a crowd of people are using the hotspot, a lot of data is consumed rather quickly.

- Don't forget to turn off the mobile hotspot when you're done using it.

Tethering the Internet connection

A more intimate way to share the phone's mobile data connection is to connect the phone directly to a computer and activate the tethering feature. Follow these steps to set up tethering:

1. **Connect the phone to a computer or laptop by using the USB cable.**

   I've had the best success with this operation when the computer is a PC running Windows.

2. **Open the Settings app and choose the More item, found in the Wireless & Networking area.**

3. **By the USB Tethering item, slide the master control to the On position.**

   On some phones, choose the Tethering item and then activate USB Tethering.

The other device should instantly recognize the phone as a "modem" with Internet access. Further configuration may be required, which depends on the computer using the tethered connection. For example, you may have to accept the installation of new software when prompted by Windows.

✔ While tethering is active, a Tethering status icon may appear, similar to what's shown in the margin.

✔ Sharing the mobile data connection incurs data usage charges against your cellular data plan. Be careful with your data usage when you're sharing a connection.

The Bluetooth Connection

Technology nerds have long had the desire to connect high-tech gizmos to one another. The Bluetooth standard was developed to sate this desire in a wireless way. Although Bluetooth is wireless *communication,* it's not the same as wireless networking. It's more about connecting peripheral devices, such as those wireless earpieces that are so popular among the golf shirt crowd.

Understanding Bluetooth

To make Bluetooth work on your phone, you need a Bluetooth peripheral, such as a wireless earpiece. The goal is to pair that gizmo with your phone. Here's how the operation works:

1. **Turn on the Bluetooth radio on both your phone and the peripheral.**

2. **Make the peripheral discoverable.**

   The peripheral broadcasts that it's available for dating other devices.

3. **On your phone, choose the peripheral from the list of Bluetooth devices.**

4. **If prompted, confirm the connection on the peripheral device.**

   For example, you may be asked to input a code or press a button.

5. **Use the Bluetooth peripheral.**

When you're done using the device, turn it off. Because the peripheral is paired with your phone, it's automatically reconnected the next time you turn it on — as long as the phone's Bluetooth radio is active.

 Bluetooth devices are marked with the Bluetooth logo, shown in the margin. It's your assurance that the gizmo can work with other Bluetooth devices.

Activating Bluetooth on the phone

You must turn on the phone's Bluetooth radio before you can use one of those Borg-earpiece implants and join the ranks of the walking connected. Assimilate these steps:

1. **Open the Settings app.**

2. **Choose Bluetooth.**

   On some Samsung phones, you'll find the Bluetooth item on the Connections tab.

3. **Ensure that the Bluetooth master control is set to the On position.**

   Slide the icon to the right to activate.

 When Bluetooth is on, the Bluetooth status icon appears. It uses the Bluetooth logo, shown in the margin.

To turn off Bluetooth, repeat the steps in this section, but slide the master control to the Off position in Step 3.

 The Bluetooth quick setting can also be used to activate and deactivate the phone's Bluetooth radio.

Pairing with a Bluetooth peripheral

To make the Bluetooth connection between your phone and some other gizmo, such as a Bluetooth headset, follow these steps:

1. **Ensure that phone's Bluetooth radio is on.**

   Refer to the preceding section.

2. **Make the Bluetooth peripheral discoverable.**

Turn on the gizmo and ensure that its Bluetooth radio is on. If the device has separate power and Bluetooth switches, press the Bluetooth button or take whatever action is necessary to make the peripheral discoverable.

3. **On the phone, open the Settings app and choose Bluetooth.**

The Bluetooth screen shows already paired and available peripherals, similar to what's shown in Figure 17-2. If not, tap the Refresh button or Scan button, or tap the Action Overflow to look for similar actions.

4. **Choose the Bluetooth peripheral from the list.**

5. **If necessary, type the device's passcode or otherwise acknowledge the connection.**

For example, press the button on the Bluetooth earpiece. That action completes the pairing.

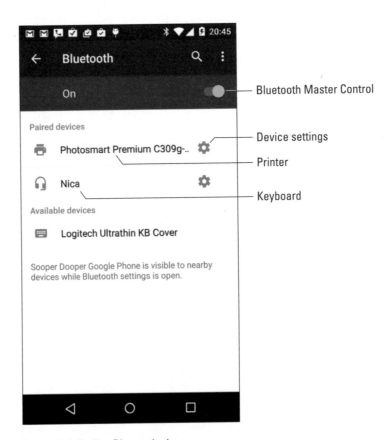

Figure 17-2: Finding Bluetooth gizmos.

After pairing, you can begin using the device.

Connected devices appear on the Bluetooth screen, under the heading Paired Devices, such as Photosmart printer and Nica earpiece (refer to Figure 17-2).

✔ To break the connection, you can either turn off the gizmo or disable the Bluetooth radio on your phone. Because the devices are paired, when you turn on Bluetooth and reactivate the device, the connection is re-established.

✔ It's rare to unpair a device. Should you need to, visit the Bluetooth screen (refer to Figure 17-2) and tap the Settings icon by the device's entry. Choose the Unpair action, or just tap the OK button to terminate the pairing.

✔ Unpair only the devices that you don't plan to use again. Otherwise, simply turn off the Bluetooth device when you're done using it.

✔ The Bluetooth radio consumes a lot of power. Don't forget to turn off the device, especially a battery-powered one, when you're no longer using it with your phone.

Connecting to your car

Another way to put Bluetooth to work is to use it for hands-free phone operation while driving. Not every car is equipped with this feature, but those that do provide yet another reason to power on the phone's Bluetooth radio and pair with a peripheral, albeit a rather large and mobile peripheral.

The specifics for pairing your phone with a Bluetooth-enabled car vary, but the general procedure works as outlined earlier, in this section "Understanding Bluetooth."

After the car and phone are paired, use the steering wheel controls to answer the phone while you're driving. Additional controls may be offered on the car's LCD screen, should it have one.

✔ Pair the phone while the car is parked. Also check for the steering wheel controls to determine which button answers the phone and which disconnects the call.

✔ I recommend plugging the phone into the car's power supply while you use it on the road.

18

Share and Store

*A*t some point, you may treasure the items inside your phone so much that you want to access them elsewhere. Or the opposite may be true, and you desire to add to your phone's inventory photos, music, or similar items. Either way, the goal is to transfer information. It seems like a simple operation, but as with most high-tech, it's a wee bit more complex than you would think.

scription ns   theater   Web Pages

From Here to There

A variety of useful and interesting methods exist for transferring files between your phone and other devices, such as a computer. None of these methods involves using tweezers or magnets.

bloggers.p df   bookmark. htm   Chapter 1.docx   E

    ✓ Getting a file into your phone is no guarantee that you can do productive things with it. Specifically, don't expect to be able to read an eBook file you've copied from elsewhere.

·'t do   dumb·

    ✓ Media you've obtained from the Google Play Store, including music, eBooks, movies, and TV shows, is accessible from any device that has Internet access. Likewise, your phone's photo and video library may be accessed remotely, as described elsewhere in this book.

✓ A good understanding of basic file operations is necessary for successful file transfers between a computer and your phone. Knowing basic procedures such as copy, move, rename, and delete, are an important part of the process. Understand how a folder works. The good news is that you don't need to manually calculate a 64-bit cyclical redundancy check on the data, nor do you need to know what a parity bit is.

Sharing files with the cloud

The simplest and most effective way to swap files between your Android phone and just about any other device is to use the *cloud.* That's just fancy talk for storing files on the Internet.

As a Google account holder, you have access to cloud storage on your Google Drive; use the Drive app to access that storage. All files saved to your Google Drive are synchronized instantly with other devices that access that storage. Change a file in a Google Drive folder and that file is instantly updated on your phone, computer — everywhere.

To move an item from a computer to the phone, first ensure that the Google Drive program is installed on the computer: Visit `drive.google.com` and follow the directions to install the program.

After the Google Drive software is installed on the computer, simply copy the file to your computer's Google Drive folder. In mere seconds, the file can be accessed on your phone by opening the Drive app.

To copy an item from the phone to your computer, follow these steps:

1. **Locate the item you want to send to the computer.**

   It can be a picture, movie, web page, YouTube video, or just about anything.

2. **Tap the Share icon.**

   If you don't see the Share icon, the item you're viewing cannot be copied to the Google Drive.

3. **Choose Drive.**

   A card appears, similar to the one shown in Figure 18-1, allowing you to view or modify how the file is shared.

4. **Edit the filename or change the folder, if you want.**

   I typically change the name to something memorable. Also, because I'm organized, I tap the Folder action bar to choose a specific Google Drive folder on which to save the item.

5. **Tap the Save icon.**

Other cloud storage apps include the popular Dropbox, Microsoft's OneDrive, the Amazon Cloud, and more. You can use those cloud storage locations just as easily as Google Drive; choose the proper app in Step 3.

Tap to edit name

Figure 18-1: Sharing a file to the Google Drive.

- ✐ All files saved to cloud storage are synchronized instantly with all devices that access the storage. Change a file on a computer and it's updated on your phone.

- ✐ You can obtain the Dropbox app from the Play Store, but you also need to get a Dropbox account and probably get the Dropbox program for the computer. Visit www.dropbox.com to get started.

- ✐ Cloud storage apps are free, and you can use a token amount of storage at no charge.

Transferring files by using the USB cable

A more direct way to transfer files between a computer and your phone is to physically connect the two devices. Because this is an ancient and traditional method of file exchange, it's also a tad bit more complex and confusing than using cloud storage to meet the same end.

To perform the direction connection file swap, obey these steps on a computer running Windows:

1. **Connect the phone and the computer by using the USB cable.**

2. **If your phone features a secure screen lock, unlock the phone.**

   The computer cannot access the phone's storage while the phone is locked with a PIN or password or another secure screen lock. See Chapter 21 for information.

3. **View the phone's storage on the computer.**

   If the AutoPlay dialog box appears, which happens most of the time, choose the Open Folder to View Files command. You see a folder window that shows files and folders on your phone.

When the AutoPlay dialog box doesn't appear, open the Computer window, and then open the Phone icon to view the device's internal storage.

4. On the PC, open the folder to which or from which you want to copy files.

Arrange both folder windows, computer and phone, on the computer screen, similar to what's shown in Figure 18-2.

Drag files to here to copy to the 'root'

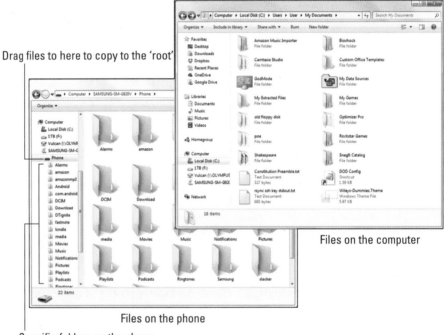

Files on the computer

Files on the phone

Specific folders on the phone

Figure 18-2: Exchanging files with an Android phone.

5. Drag the file icon(s) from one folder window to the other.

Dragging the file copies it.

6. Close the folder windows when you're done copying files.

7. Disconnect the USB cable from the computer.

If you want to be specific, drag the file to the phone's download folder or to the root folder (refer to Figure 18-2). On the PC, drag it to the My Documents folder. You can also drag files directly to the desktop and decide later where to store them.

On the odd chance that you use an Android phone with a Macintosh, you must obtain special software to perform a USB cable file transfer. Obtain the Android File Transfer program from this website:

```
www.android.com/filetransfer
```

Install the software. Run it. From that point on, when you connect your Android phone to the Macintosh, a special window appears, similar to the one shown in Figure 18-3. It lists the phone's folders and files. Use that window for file management, as covered in this section.

Figure 18-3: The Android File Transfer program.

- ✔ If you're transferring media, such as music or photos, connect the phone to the PC by using the USB cable, but run the Windows Media Player program in Windows. See Chapter 14 for information.

- ✔ If your phone features removable storage, that media also appears in the Phones folder in Windows or as a separate button in the Android File Transfer program on a Macintosh.

- ✔ Files you've downloaded on the phone are stored in the Download folder.

- ✔ Pictures and videos on the phone are stored in the DCIM/Camera folder.

- ✔ Music on the phone is stored in the Music folder, organized by artist.

Printing

You may not think of it as part of moving files back and forth, but using a printer with your Android phone is a type of file transfer. It just so happens that the other device is a printer, not a computer.

Media card transfer

If your phone features removable storage, you can use it to transfer files. Remove the MicroSD card from the phone and insert it into a computer. From that point, the computer can read the files just as they can be read from any media card.

See Chapter 1 for details on how to remove the MicroSD card from your phone. You can't just yank out the thing! You also need a MicroSD adapter to insert the card into a media reader on the computer. Or you can get a MicroSD card USB thumb drive adapter.

Android printing works like this:

1. **View the material you want to print.**

   You can print a web page, photo, map, or any number of items.

2. **Tap the Action Overflow.**

   Some Samsung phones use the MORE button in place of an Action Overflow icon.

3. **Choose Print.**

   If you don't see the Print action, your phone lacks this feature.

4. **Select a printer.**

   The current printer is shown on the action bar, as illustrated in Figure 18-4. If that's the printer you want to use, you're good. Otherwise, tap the action bar to view printers available on the currently connected Wi-Fi network.

5. **To change any print settings, tap the Show More Details chevron.**

   The items presented let you set which pages you want to print, change the number of copies, and make other common printer settings.

6. **Tap the Print button.**

   The item prints.

Not every app supports printing. If you work through Steps 1 and 2 in this section and don't find the Print action, the feature might not be available. Check the Settings app to be sure: Locate a Printing item and ensure that all printing services are activated. On some Samsung phones, the Printing item is found by choosing the MORE item in the Settings app's Wireless & Networks section.

Screencasting media

When you desire to watch movies, look at your photos, or listen to music on a large-format screen, it's time to *screencast:* This technology takes the image presented on your phone (or music playing) and casts it to an HDTV or monitor.

To successfully cast media, the HDTV or monitor must have a screencasting gizmo installed, such as Google's Chromecast. After you set up that gizmo, the screencast works like this:

1. **Open the app that plays the media you want to watch or listen to.**

   Compatible apps include Play Music, Play Movies & TV, YouTube, Netflix, and similar apps.

Show more details

Select printer Action Bar   Print

Print preview

Figure 18-4: Android phone printing.

2. **Ensure that the HDTV or monitor is set to the proper HDMI input.**

   For example, if a Chromecast dongle is installed on HDMI Input 4, switch the TV to that input. The casting dongle must be awake and active.

3. **Tap the Chromecast icon.**

   The icon appears similar to the one shown in the margin. If you don't see this icon, the media cannot be cast to another device or the casting gizmo isn't on (refer to Step 2).

4. **Choose a screencasting device from the list.**

   The phone begins to send the media to the other device.

You can still use your phone while it's casting. The app may offer you controls, such as Play and Pause, or it might display additional information about the media.

To end a screencasting session, tap the Chromecast icon again and tap the Disconnect button.

Phone Storage Mysteries

Somewhere deep in your phone's bosom lies a storage device. It's like the hard drive in a computer: The thing can't be removed, but that's not the point. The point is that the storage is used for your apps, music, videos, pictures, and a host of other information. This section describes what you can do to manage that storage.

✔ Android phones come with 8GB, 16GB, 32GB, or more internal storage.

✔ Removable storage in the form of a MicroSD card is available on some phones, although this option is becoming less popular. The MicroSD card capacity varies between 8GB and 64GB or greater.

✔ A GB is a *gigabyte,* or 1 billion bytes (characters) of storage. A typical 2-hour movie occupies about 4GB of storage. Most things you store on the phone don't occupy that much space. In fact, media stored on the cloud occupies hardly any storage space. See Chapter 16 for details on how media is stored on the cloud and can be downloaded to your phone.

Reviewing storage stats

To see how much storage space is available on your phone, follow these steps:

1. **Open the Settings app.**

2. **Choose the Storage item.**

   On Samsung phones, you'll find the Storage item on the Settings app's General tab.

A typical storage screen is shown in Figure 18-5. It details information about storage space on the phone's internal storage and, if available, on the MicroSD card.

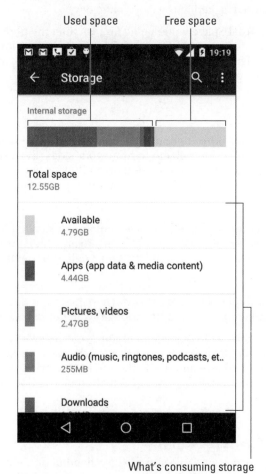

Figure 18-5: Phone storage information.

If your phone has external storage, look for the SD Card category at the bottom of the Storage screen (not shown in Figure 18-5).

Tap a category on the Storage screen to view details on how the storage is used or to launch an associated app. For example, tap Apps to view a list of running apps. If you tap the Pictures, Videos item you can view pictures and videos.

✓ Videos, music, and pictures (in that order) consume the most storage space on your phone.

✓ Media that you download to the phone from the Google cloud, such as movies or videos, are tracked in the Apps category.

✓ Don't bemoan the Total Space value being far less than the phone's storage capacity. For example, in Figure 18-5, a 16GB phone shows only 12.55 GB available. The missing space is considered overhead, as are several gigabytes taken by the government for tax purposes.

Managing files

You probably didn't get an Android phone because you enjoy managing files on a computer and wanted another gizmo to hone your skills. Even so, you can practice the same type of file manipulation on a phone as you would on a computer. Is there a need to do so? Of course not! But if you want to get dirty with files, you can.

Some Android phones come with a file management app. It's called My Files or File Manager, and it's a traditional type of file manager, which means that if you detest managing files on your computer, you'll enjoy the same pain and frustration on your phone.

✓ If you simply want to peruse files you've downloaded from the Internet, open the Downloads app, found in the apps drawer. Refer to Chapter 10.

✓ When your phone lacks a file management app, you can swiftly obtain one. An abundance of file management apps are available at the Google Play Store. One that I admire and use is the ASTRO file manager/browser, from Metago. See Chapter 16 for more information on the Google Play Store.

Unmounting the MicroSD card

The MicroSD card provides removable storage on a handful of Android phones. When the phone is turned off, you can insert or remove the MicroSD card at will; directions are provided in Chapter 1. The MicroSD card can also be removed when the phone is turned on, but you must first unmount the card. Obey these steps:

1. **Open the Settings app.**

2. **Choose Storage.**

   On some Samsung phones, tap the General tab to locate the Storage item.

3. **Choose Unmount SD Card.**

   This item is found near the bottom of the screen. If not, your phone lacks the ability to host a removable MicroSD card.

4. **Ignore the warning and tap the OK button.**

5. **When you see the Mount SD Card action, it's safe to remove the MicroSD card.**

It's important that you follow these steps to safely remove the MicroSD card. If you don't and you simply pop out the card, it can damage the card and lose information.

You can insert a MicroSD card at any time. See Chapter 1 for details.

Formatting MicroSD storage

Your Android phone is designed to instantly recognize a MicroSD card when it's inserted. If it doesn't, you can attempt to format the card to see whether that fixes the problem.

All data on the MicroSD card is erased by the formatting process.

To format a MicroSD card, first insert it into the phone per the directions found in Chapter 1. Next, follow these steps to determine whether a format is in order:

1. **Open the Settings app.**

2. **Choose the Storage item.**

3. **Choose Format SD Card.**

4. **Tap the Format SD Card button.**

5. **Tap the Delete All button to confirm.**

   The MicroSD card is unmounted, formatted, and then mounted again and made ready for use.

After the card is formatted, you can use it to store information, music, apps, photos, and stuff like that.

The Apps-and-Widgets Chapter

In This Chapter

▶ Organizing the Home screen

▶ Working with widgets

▶ Rearranging the Home screen

▶ Creating folders

▶ Performing app updates

▶ Removing apps

▶ Selecting default apps

▶ Organizing the Apps drawer

*B*eyond making phone calls, your Android phone is host to an abundance of apps. The app launchers, as well as their widget cousins, adorn your phone's Home screen, which helps make the device more usable. Your goal is to keep these apps and widgets accessible without overwhelming the Home screen.

Apps and Widgets on the Home Screen

Your phone came with app launchers and widgets affixed to the Home screen. You can keep those launchers, add your own, remove some, and change their locations. And when the Home screen brims with launchers like an overflowing teacup, you can build folders to further organize and keep your apps neat and tidy.

The app icons on the Home screen are called *launchers*.

Adding launchers to the Home screen

You need not live with the unbearable proposition that you're stuck with only the apps that come preset on the Home screen. Nope — you're free to add your own apps. Just follow these steps:

1. **Visit the Home screen page on which you want to stick the launcher.**

   The screen must have room for the icon. If not, swipe the screen left or right to hunt down a page.

2. **Tap the Apps icon to display the apps drawer.**

3. **Long-press the app you want to add to the Home screen.**

   After a moment, the Home screen page you chose in Step 1 appears, similar to what's shown in Figure 19-1.

4. **Drag the app to a position on the Home screen.**

   Launchers are aligned to a grid. Other launchers may wiggle and jiggle as you find a spot. That's okay.

5. **Lift your finger to place the app.**

The app hasn't moved: What you see is a copy or shortcut. You can still find the app in the apps drawer, but now the app is also available — more conveniently — on the Home screen.

- Any new apps you obtain from the Play Store are automatically affixed to the Home screen. If you prefer not to have the new app's launcher on the Home screen, you can remove it. See the later section "Evicting items from the Home screen."

- Not every app needs a launcher icon on the Home screen. I recommend placing only those apps you use most frequently.

- You can't cram more launchers on the Home screen than will fit in the grid. As an alternative, consider rearranging launchers by moving some to another Home screen page, as covered in the later section "Moving launchers and widgets." Also see the section "Building folders" for another solution to a crowded Home screen. Finally, if your phone allows you to add more Home screen pages, you can solve the problem that way; see Chapter 20.

Putting a launcher on the favorites tray

The line of launchers at the bottom of the Home screen is called the *favorites tray*. It includes the Apps icon, as well as other launchers you place there, such as the Phone or Dialer launcher, or any app you use most often. The favorites tray launchers appear at the bottom of every Home screen page.

Drag the icon to a position on the
Home screen

Favorites tray

Figure 19-1: Placing a launcher icon on the Home
screen.

Launchers are added to the favorites tray in one of two ways:

- **Drag a launcher off the favorites tray,** either to the Home screen or to the Remove or Delete icon. You can then add another launcher to replace the one you removed.

- **Drag a launcher from the Home screen to the favorites tray,** in which case any existing launcher swaps places with the one already there.

Of these two methods, the second one may not work on all phones. In fact, the second method may create a folder in the favorites tray, which is probably not what you want.

See the later section "Building folders" for information on creating and using Home screen folders, which can also dwell on the favorites tray.

 The best icons to place on the Home screen are those that show updates, such as new messages, similar to the icon shown in the margin. These icons are also ideal to place in the favorites tray, as covered in the next section.

Slapping down widgets

The Home screen is the place where you can find *widgets,* or tiny, interactive information windows. A widget often provides a gateway to a specific app or displays information such as status updates, the name of the song that's playing, or the current weather.

Two common ways exist to add widgets to the Home screen. The stock Android method is to view the Widgets drawer. An older method involved accessing widgets from a tab on the apps drawer. Both methods are covered in these steps:

1. **Switch to a Home screen page that has enough room for the new widget.**

   Widgets come in a variety of sizes. The size is measured by launcher dimensions: A 1 x 1 widget occupies the same space as a launcher icon. A 2 x 2 widget is twice as tall and twice as wide as a launcher icon.

2. **Access the widgets.**

   On newer phones, long-press a blank part of the Home screen. Tap the Widgets icon.

   If you don't see the Widgets icon, tap the Back navigation icon, and then tap the Apps icon to view the apps drawer. Tap the Widgets tab atop the screen.

3. **Long-press the widget you want to add.**

   Swipe through the pages to find widgets, which are listed alphabetically and show preview images just like apps. Also shown are the widget's dimensions.

4. **Drag the widget to the Home screen.**

   Move the widget around to position it. As you drag the widget, existing launchers and widgets jiggle to make room.

5. **Lift your finger.**

   If the widget grows a border, it can be resized. See the next section.

You may see a prompt for additional information after adding certain widgets — for example, a location for a weather widget or a contact name for a contact widget.

- ✔ The variety of available widgets depends on the apps installed. Some apps come with widgets; some don't. Some widgets are independent of any app.

- ✔ Just like launchers, widgets can be moved and rearranged on the Home screen. See the later section "Moving launchers and widgets," for obtaining the proper *feng shui*.

- ✔ Also see the later section "Evicting items from the Home screen," for information on unsticking a widget from the Home screen.

Resizing a widget

Some widgets are of a fixed size, but many can be resized. You can resize a widget immediately after plopping it down on the Home screen, or at any time, really: The secret is to long-press the widget: If it grows a box, as shown in Figure 19-2, you can change the widget's dimensions.

Figure 19-2: Resizing a widget.

To resize, drag one of the resizing dots in or out, as illustrated in Figure 19-2. Tap elsewhere on the touchscreen when you're done resizing.

Moving launchers and widgets

Redecorating your Android phone's Home screen is a lot easier than trying to move a couch from one room in your house to another. And you don't have to bribe anyone with pizza to make it happen: Simply long-press the item, as illustrated in Figure 19-3.

Figure 19-3: Moving an item on the Home screen.

Drag the launcher, widget, or folder to another position on the Home screen. If you drag to the far left or far right of the screen, the item is sent to another Home screen page.

Evicting items from the Home screen

 When your rearranging plans include the removal of a launcher, widget, or folder, drag that item up to the top of the Home screen page, where the Remove icon lurks. The Remove icon can appear as shown in the margin, or it may be a giant X or the word *Remove*. When the item hovers over that part of the screen, it turns red. Lift your finger and it's gone.

✔ Feel no guilt or regret about removing anything from the Home screen, especially those launchers and widgets preinstalled by the phone's manufacturer or the cellular provider. If you don't use something on the Home screen, get rid of it.

✔ Removing a launcher from the Home screen doesn't uninstall the app. See the later section "Uninstalling apps."

Building folders

To further organize the Home screen, or to deal with an overly crowded Home screen, consider gathering similar apps into folders. A *folder* is simply a single icon that contains a host of launchers. For example, I have a Listen folder that contains all my streaming-music apps. Your phone may have come with a Google folder, which contains all of Google's various apps.

The stock Android method to create a folder is to drag one launcher over another on the Home screen. Be quick! When the two launchers get close, a circle encloses them, which provides visual feedback that a folder is created.

Some phones require that you first create a folder: Drag a launcher to a Folder icon or Create Folder icon. Or you may have to first tap the Create Folder icon, name the folder, and then add launchers.

Figure 19-4: Folder icon varieties.

Figure 19-4 illustrates different ways that folders appear on various phones. The two most common presentations are to show app icons inside of a circle or pasted to a folder icon.

Continue to build a folder by dragging in more launchers.

To use a folder, tap its icon. The folder opens, displaying its contents. Tap a launcher to start the associated app. Or if you don't find what you want, tap the Back navigation icon to close the folder.

✔ Folders are managed just like other items on the Home screen. You can rearrange folder icons and remove them just like launchers and widgets. Removing a folder doesn't uninstall its enclosed aps.

✔ Change a folder's name by opening the folder and then tapping its name. Use the onscreen keyboard to type a new name.

✔ To remove a launcher from a folder, open the folder and drag out the icon. When the second-to-last icon is dragged out of a folder, the folder is removed. If not, drag the last icon out, and then remove the folder as described in the earlier section "Evicting items from the Home screen."

Manage Those Apps

The apps you install on your Android phone originate from the Play Store. And that's where you can return for app management. The task includes reviewing apps you've downloaded, updating apps, and removing apps that you no longer want or severely hate.

Reviewing your apps

To peruse the apps obtained from the Google Play Store, follow these steps:

1. **Open the Play Store app.**

2. **Choose My Apps from the navigation drawer.**

    Tap the Side Menu icon, as shown in the margin, to view the navigation drawer.

3. **Peruse your apps.**

Apps are presented in two categories, Installed and All, as shown in Figure 19-5. Apps currently installed on your phone are listed on the Installed tab. The All tab lists every app you've ever obtained from the Play Store, whether or not it's currently installed on the phone.

See the next section for information on using the Update All button, shown in Figure 19-5.

✔ Tap an app in the list to view its details. Some of the options and settings on the app's details screen are discussed elsewhere in this chapter.

✔ Uninstalled apps remain on the All list because you did, at one time, download the app. To reinstall an app (and without paying a second time for paid apps), choose the app from the All list and tap the Install button.

Updating apps

New versions, or updates, of apps happen all the time. It's nothing you need to worry about, because most app updates are installed automatically. Occasionally, however, you must tend to manual updates.

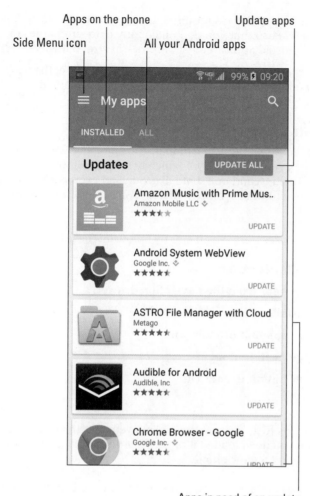

Figure 19-5: Apps on your phone.

 When apps are in need of update approval, an App Update notification appears, similar to the one shown in the margin. Choose that notification, or visit the My Apps screen as described in the preceding section (and shown in Figure 19-5). Tap the Update or Update All button. You may have to accept permissions for some or all of the apps: Tap the Accept button when prompted.

You can view the app update process in the Play Store app or go off and do something else with your phone.

 ✔ Most apps update automatically; you need not do a thing. The updates generate a notification icon, shown in the margin. Feel free to dismiss that notification.

✔ An Internet connection is required in order to update apps. If possible, try to use Wi-Fi so that you don't incur any data surcharges on your cellular bill. Android apps aren't super-huge in size, but why take the risk?

✔ If the Internet connection is broken during an update, the updates automatically continue when the connection is reestablished.

Uninstalling apps

It doesn't happen often, but you can uninstall apps that you no longer use or those that vex you incessantly. I typically uninstall apps after downloading a clutch of them to find one that I like. Such an exercise isn't required, but it helps keep me sane, which is an item somewhere toward the top of my priority list.

When the need arises, remove an app by following these steps:

1. **Open the Play Store app.**

2. **Choose My Apps from the navigation drawer.**

3. **Tap the Installed tap.**

4. **Choose the app that offends you.**

5. **Tap the Uninstall button.**

6. **Tap the OK button to confirm.**

   The app is removed.

The app continues to appear on the All list even after it's been removed. That's because you downloaded it once. That doesn't mean that the app is currently installed on your phone.

✔ You can always reinstall paid apps that you've uninstalled. You aren't charged twice for doing so.

✔ You can't remove apps that are preinstalled on the phone by either the phone's manufacturer or cellular provider. I'm sure there's a technical way to uninstall the apps, but seriously: Just don't use the apps if you want to remove them and discover that you can't.

Choosing a default app

While tapping links or items or attempting to view a download, you may find yourself confronted with the Open With card, which looks similar to the one shown in Figure 19-6. Your duty is to choose from one or more apps to complete an action. Then you're given the choice of the options Just Once or Always.

✔ When you choose Always, the chosen app is always used to complete the action; you'll never see the Open With card again for whatever task you're performing.

✔ When you choose Just Once, you see the Open With card again.

My advice is to choose Just Once until you get sick of seeing the Open With card. At that point, choose Always.

The fear, of course, is that you'll make a mistake. Keep reading in the next section.

The Open With card might be titled Complete Action Using on some phones.

Clearing default apps

Fret not, gentle reader. The action you chose for the Open With card can be undone. For example, if you select the Amazon Kindle app shown in Figure 19-6, you can undo that choice by following these steps:

1. **Open the Settings app.**

2. **Choose Apps or Applications.**

   Some Samsung phones may feature an Applications tab, in which case you choose that tab to continue.

3. **Select the app that always opens.**

   This is the tough step because you must remember which app you chose to "Always" open.

   If your Samsung phone offers a Default Applications item on the Applications screen, choose it to quickly locate default apps.

4. **Tap the Clear Defaults button.**

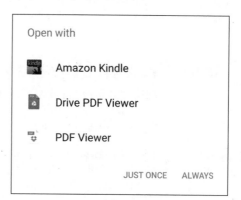

Figure 19-6: The Open With question is posed.

After you clear the defaults for an app, you see the Open With card again. The next time you see it, however, make a better choice.

Shutting down an app run amok

Sometimes an app goofs up or crashes. You may see a warning card on the touchscreen, informing you that the app has been shut down. That's good.

What's better is that you too can shut down apps that misbehave or those you cannot otherwise stop. Follow these steps:

1. **Open the Settings app.**

2. **Tap the Apps or Applications item.**

   On Samsung phones, choose the Application Manager item. You might need to tap an Applications tab or General tab to find that item.

3. **Tap the Running tab.**

4. **Choose the errant app from the list.**

   For example, if the Annoying Sound app is bothering you, tap it from the list.

5. **Tap the Stop or Force Stop button.**

   The app is terminated.

Use these steps only as a last-ditch, drastic act. Don't kill off any app or service unless the app is annoying or you are otherwise unable to stop it. Avoid killing off Google Services, which can alter the phone's behavior.

Apps Drawer Organization

Some phones offer tools for arranging apps in the apps drawer. These tools allow you to present the apps in an order other than alphabetical, rearrange the apps, or even collect apps and place them into folders.

The key to organizing the apps drawer is to look for the Action Overflow or Edit icon. If it's available, tap that icon and peruse available options for modifying the Apps drawer presentation. Because this is a nonstandard Android feature, each phone implements Apps drawer organization differently.

Your phone might also allow you to create folders on the Apps drawer. Use these folders to collect and organize apps in one location, similar to how folders work on the Home screen.

As with organizing the Apps drawer, tap the Action Overflow or Edit icon to look for a Create Folder command. Generally speaking, folders on the App drawer work like folders on the Home screen: Open the folder to view icons, drag an icon into or out of the folder, and so on. Refer to the section "Building folders," earlier in this chapter, for details.

20

Customize and Configure

Customizing your Android phone doesn't involve sprucing up the phone's case, so put away that Bedazzler™. The kind of customization this chapter refers to involves fine-tuning the way the Android operating system presents itself. You can modify the Home screen, adjust the display, customize the keyboard, and change sounds. The goal is to truly make your phone your own.

It's Your Home Screen

The typical Android Home screen sports anywhere from three to nine pages and a specific background, or *wallpaper,* preset by the phone's manufacturer. Or, if the phone has a preset wallpaper touting your cellular provider's new 4G LTE service, I'd guess that they provided the wallpaper. No matter how the Home screen is presented, it's yours to customize.

Finding the Home screen menu

To view the Home screen's menu, long-press a blank part of the Home screen; do not long-press a launcher or widget. The menu actions may present themselves as icons or as a list of actions on a card. Both presentations are shown in Figure 20-1.

Home screen management icons

Home screen management menu

Figure 20-1: The Home screen menus.

The Home screen menu includes some or all of the following actions:

- ✔ **Wallpapers:** Change the background image on the Home screen. This item might be titled Wallpaper or Set Wallpaper.

- ✔ **Widgets:** Choose a new widget to affix to the Home screen. This item might be titled Apps and Widgets.

- ✔ **Settings:** Start the Setting apps.

- ✔ **Folder:** Create folders for multiple apps on the Home screen.

- ✔ **Page:** Add, remove, or manage multiple Home screen pages.

More actions might be available as well, depending on the phone. Generally speaking, the Wallpapers and Widgets items are the most popular.

- ✔ You cannot long-press the Home screen when it's full of launchers and widgets. Swipe to a screen that's all or partially empty.

✔ On some older Android phones, the only action available is to change the Home screen wallpaper.

Changing wallpaper

The Home screen background can be draped with two types of wallpaper: traditional and live. Traditional wallpaper can be any image, such as a picture you've taken or an image provided by the phone's manufacturer. Live wallpaper is animated or interactive.

To hang new wallpaper on the Home screen, obey these steps:

1. **Long-press the Home screen.**

2. **Choose Wallpapers or Set Wallpaper.**

   Refer to Figure 20-1 to see how the actions could be presented.

3. **Tap a wallpaper to see a preview.**

   Swipe the list left or right to peruse your options. You see the previous wallpaper images plus those provided by the phone's manufacturer. On the far right, you'll find the live wallpapers.

4. **Tap the Set Wallpaper button to confirm your choice.**

   The new wallpaper takes over the Home screen.

Some phones may prompt you to set the Home screen wallpaper, Lock screen, or both.

To select an image stored in the phone, or one you've taken by using the phone's camera, choose the item Pick Image or From Gallery. Select an image. You may be prompted to crop the image so that it fits properly on the Home screen.

You might be prompted to choose a wallpaper type before you can view samples. The list includes Wallpapers (preset images), Live Wallpapers, and Gallery or Photos, which represents images you've taken.

✔ Be careful how you crop the wallpaper image when you choose one of your own photos. Zoom out (pinch your fingers on the touchscreen) to ensure that the entire image is cropped properly for both the phone's horizontal and vertical orientations.

✔ Live wallpapers can be obtained from the Google Play Store. See Chapter 16.

✔ The Zedge app offers some interesting wallpapers. Check it out at the Google Play Store.

✔ See Chapter 13 for more information about the images stored on your phone, including information on how to crop an image.

Managing Home screen pages

The number of pages on the Home screen isn't fixed. You can add pages. You can remove pages. You can even rearrange pages. This feature might not be available to all Android phones and, sadly, it's not implemented in exactly the same way.

The stock Android method of adding a Home screen page is to drag a launcher left or right, just as though you were positioning that item on another Home screen page. When a page to the left or right doesn't exist, the phone automatically adds a new, blank page.

Likewise, to remove a page, simply drag off the last launcher. The page vanishes.

Other phones may be more specific in how pages are added. For example, you might be able to choose a Page command from the Home Screen menu (refer to Figure 20-1).

Samsung phones feature a Home screen page overview, shown in Figure 20-2. To edit Home screen pages, pinch the Home screen. On older Samsung phones, long-press the Home screen and choose Edit Page. You can then manage Home screen pages as illustrated in the figure.

Generally speaking, to rearrange the pages, long-press a page and drag it to a new spot. When you're done, tap the Back navigation icon.

 ✔ The total number of Home screen pages is fixed. The maximum may be three, five, seven, or nine, depending on your phone.

 ✔ On some phones, the far left Home screen page is the Google Now app.

 ✔ Some phones allow you to set the primary Home screen page, which doesn't necessarily have to be the center Home screen page. I've seen different ways to accomplish this task. The most common one is to tap the Home icon in a thumbnail's preview, which is what's illustrated in Figure 20-2.

Lock Screen Settings

The Lock screen is different from the Home screen, although the two locations share similar traits. As with the Home screen, you can customize the Lock screen. You can change the background, add app launcher shortcuts, and do all sorts of tricks.

 ✔ Lock screen features might not be available for all Android phones.

 ✔ For information on setting the screen lock, as well as controlling the Lock screen notifications, refer to Chapter 21.

Drag here to remove a page

Main Home screen page

Drag pages to rearrange

Add page

Wallpapers   Widgets   Themes   Screen grid

Figure 20-2: Manipulating Home screen pages.

Setting the Lock screen wallpaper

For most Android phones, the Lock screen wallpaper is the same as the Home screen wallpaper. A few phones, however, let you set separate Lock screen wallpaper.

To determine whether your phone provides for separate Lock screen wallpaper, follow the same steps as outlined in the earlier section "Changing wallpaper." If you see a prompt to set the Lock screen wallpaper, choose it. Then select for the Lock screen a wallpaper that's different from the one on the Home screen.

✔ Look for a prompt on the Set Wallpaper screen to see whether you can set the Lock screen wallpaper or both Lock and Home screen wallpaper at the same time.

✔ I've seen phones that feature a Lock Screen item in the Settings app. Choose that item to set the Lock screen wallpaper.

Adding Lock screen shortcuts

The current trend is for the phone's Lock screen to feature one or two shortcuts. These shortcuts let you quickly access the Phone app and Dialer app, often without unlocking the device: Simply swipe the shortcut icon, and the associated app starts.

Your phone may allow you to add more Lock screen shortcuts. For example, you can place icons for your favorite apps on the Lock screen and then swipe those launchers to unlock the phone and run the associated app.

If your phone offers the Lock screen shortcut feature, follow these steps:

1. **Open the Settings app.**

2. **Locate the Lock Screen item.**

   If your phone lacks this item, customizing Lock screen shortcuts isn't a feature.

   On some Samsung phones, look for the Lock Screen item on the Device tab.

3. **Choose the Shortcuts item.**

   This item shows up only when a nonsecure screen lock is set.

4. **Select the Lock screen shortcuts.**

   At this point, the specific steps vary from phone to phone.

After activating the Shortcuts feature and choosing which apps to use, lock your phone and then unlock it to review your choices.

Lock screen shortcuts work best for the swipe screen lock. See Chapter 21 for information on screen locks.

Display Settings

The Display item in the Settings app deals with touchscreen settings. Two popular items worthy of your attention are the Brightness and Screen Timeout options.

The Display item might be titled Display and Wallpaper on your Samsung phone. It might be found under the Device tab.

Adjusting display brightness

The phone's touchscreen can be too bright, too dim, or just right. Which setting is best? That's up to you. Follow these steps:

1. **Open the Settings app and tap the Display item.**

2. **Tap the Brightness item.**

   This item might be titled Brightness Level.

3. **Adjust the slider to set the touchscreen's intensity.**

To have the phone adjust the brightness for you, locate the Adaptive Brightness item and slide its master control to the On position. Not every phone may feature this option.

> ✔ The Adaptive Brightness setting might be called Automatic Brightness. In the Quick Settings area, it might appear as the Auto check box.

> ✔ You might also find a Brightness setting in the phone's Quick Settings. See Chapter 3 for information on accessing the Quick Settings.

Controlling the screen lock timeout

Your Android phone automatically locks itself after a given period of inactivity. For example, after 1 minute of your not touching the phone, the screen dims and then the phone locks. You can control that duration by following these steps:

1. **Open the Settings app and choose Display.**

2. **Choose the Screen Timeout item.**

   This item might be titled Sleep on some phones.

3. **Set a timeout value from the list.**

   The standard value is 1 minute.

The screen lock timeout measures inactivity. When you don't touch the screen or press a key, the timer starts ticking. A few seconds before the timeout value you set (refer to Step 3), the touchscreen dims. Then it locks. If you touch the screen before then, the sleep timer is reset.

Keyboard Settings

The standard Android keyboard is called the *Google Keyboard*. It offers some special features that supposedly make the onscreen typing experience more enjoyable. I'll leave it up to you to determine whether that's true.

Keyboard settings are held in the Settings app, accessed by tapping the Language & Input item. On some Samsung phones, that item is found on either the General tab or Controls tab in the Settings app.

Generating keyboard feedback

Typing seems to work better with an onscreen keyboard when the Haptic Feedback feature is active. To check this setting, follow these steps:

1. **Open the Settings app and choose Language & Input.**

2. **Choose Google Keyboard and then Preferences.**

   On Samsung phones, this item is titled Samsung Keyboard. It's the same thing. There is no need to choose Preferences for the Samsung keyboard.

3. **Use the master control to set the vibration and sound options.**

Activating predictive text

The Google Keyboard enables its predictive text automatically, but not every Android phone uses that keyboard. To ensure that the feature is active, follow these steps:

1. **Open the Settings app and choose Language & Input.**

2. **Choose Google Keyboard and then Text Correction.**

   Some Samsung phones show the Samsung Keyboard item. Choose it to view the Predictive Text setting.

3. **Ensure that the item Next-Word Suggestions or Predictive Text is enabled.**

   This item may instead be titled Show Suggestions.

If you can't locate a Next-Word Suggestions or Predictive Text item, it's most likely on all the time and can't be disabled.

See Chapter 4 for more information on using the Predictive Text feature.

Activating gesture typing

Gesture typing allows you to create words by swiping your finger over the onscreen keyboard. Chapter 4 explains the details, although this feature may not be active on your phone. To ensure that it is, follow these steps:

1. **Open the Settings app and choose Language & Input.**

2. **Choose Google Keyboard, and then choose Gesture Typing.**

   On a Samsung phone, choose Samsung Keyboard and then Keyboard Swipe.

3. **Ensure that all Master Control icons are set to the On position.**

   Only the Enable Gesture Typing item needs to be enabled, although activating the other items enhances the experience.

Samsung phones show the item titled SwiftKey Flow, which is essentially the same thing as gesture typing.

Turning on dictation

Voice input should be enabled automatically on your Android phone, although that's no guarantee. The secret is to find the Dictation (Microphone) icon on the keyboard. The icon looks similar to the one shown in the margin.

On some Samsung keyboards, long-press the Multifunction key to locate the Dictation icon.

When you can find the Dictation key, and before you toss the phone into the garbage disposal, follow these steps to ensure that this feature is active:

1. **Open the Settings app and choose Language & Input.**

2. **Choose Google Voice Typing.**

3. **Ensure that the feature is enabled.**

   On many phones, the feature can't be disabled, so after choosing Google Voice Typing (refer to Step 2), you see a list of features.

Information on disabling the voice-typing word filter is found in Chapter 24.

Audio Adjustments

Yes, the phone makes noise. Incoming calls ring, you hear the sound of someone else on the line, and the various apps, alarms, and media beep, bleep, and blort. The Settings app is the place to go when the phone's sound needs fine-tuning. Tap the Sound & Notification item in the Settings app to make those various audio adjustments.

Setting the volume

To adjust the volume for all potential sounds the phone can make, follow these steps:

1. **Open the Settings app and choose Sound & Notification.**

   This item might be titled Sound on some phones, or Sounds and Notifications.

2. **Choose the Volume item.**

   Choosing this item is necessary only when you don't directly see the sound adjustment sliders.

3. **Adjust the sliders to set the volume for various noises the phone makes.**

   Slide to the left to make a sound quieter; slide to the right to make a sound louder. When you lift your finger, you hear a sound preview.

The Ring or Ringtone Volume sets the phone's ringtone volume. This category includes incoming calls and notifications, although some phones may feature a separate slider for notifications.

The media volume controls the sound for movies, videos, audio in the web browser, and so on.

The alarm volume sets the intensity used for the Clock app's alarm.

System appears on some phones as a category for general Android sounds.

If you'd like the phone to vibrate on an incoming call, enable the Also Vibrate for Calls setting. This item might be titled Vibrate When Ringing, and it may be found on a separate Vibrations item on the Sounds and Notifications screen.

The volume can also be set at any time by using the Volume key on the side of the phone. The Volume key adjusts the sound loud (up) or soft (down) for whatever you're listening to, such as music or an incoming call. You can adjust the volume by using the Volume key whether or not the phone is locked.

Selecting the phone's ringtone

To set a new ringtone for your phone, or to simply confirm which ringtone it's using, follow these steps:

1. **Open the Settings app and choose Sound & Notification.**

2. **Choose Phone Ringtone.**

   The item might be titled Ringtones and Sounds or just Ringtone.

   If a ringtone app is installed, you see the Complete Action Using card, prompting you to choose a ringtone app or source. For now, choose Media Storage or Android System, and then tap the Just Once button.

3. **Choose a ringtone from the list.**

   You hear the ringtone's preview.

4. **Tap OK to set the new ringtone, or tap the Back Navigation icon.**

   To keep the current ringtone, tap Cancel or tap the Back navigation icon without choosing a new ringtone.

The phone's notification ringtone is set separately from the incoming-call ringtone. To set the notification ringtone, choose the item Default Notification Ringtone in Step 2.

✔ Various apps may set their own ringtones, such as text messaging ring-tones, and alert sounds for Facebook. These ringtones are set within the given app: Look for a Settings action in the app, either found on the navigation drawer or accessed by tapping the Action Overflow icon.

✔ To disable the ringtone, choose None in Step 3. If you do so, I recommend setting the vibration so that the phone vibrates on an incoming call. Further, keep in mind that it's possible to temporarily disable sound on your phone. See Chapter 3.

✔ Vibration settings are also found on the Sound & Notification screen in the Settings app.

✔ Some phones may let you set your own ringtones. If the phone features a Music app (not Play Music), tap the music or sound's Action Overflow icon and choose Set As Ringtone.

✔ Refer to Chapter 19 for information on the Complete Action Using card, which falls under the category of setting a default app.

Setting a contact's ringtone

Ringtones can be assigned by contact so that when your annoying friend Larry calls, you can have your phone yelp like a whiny puppy. Here's how to set a ringtone for a contact:

1. **Open the Contacts app.**

2. **Choose the contact for whom you want to assign a ringtone.**

3. **Edit the contact.**

   Tap the Edit icon (shown in the margin) or tap the Edit button.

4. **Choose Ringtone.**

   When a Ringtone item isn't directly visible, tap the Action Overflow and look for it there. It might be titled Set Ringtone.

   If prompted with a Complete Action Using card, choose Media Storage or Android System and then tap Just Once.

5. **Choose a ringtone.**

   The ringtone you choose plays a brief preview.

6. **Set the ringtone.**

   On some phones, tap the OK button to assign the ringtone to the contact. On other phones, choosing a ringtone from the list (Step 5) sets the ringtone automatically. In that case, tap the Back navigation icon and you're done.

Whenever the contact calls, the phone rings using the ringtone you've specified.

To remove a specific ringtone for a contact, repeat the steps in this section, but in Step 5 choose Default Ringtone. (It's found at the top of the list.) This choice sets the contact's ringtone to be the same as the phone's ringtone.

Also see Chapter 7 for information on banishing a specific contact directly to voice mail.

Android Phone Security

As more and more of your life is surrendered to the digital realm, the topic of security grows in importance. This concern extends directly to your phone, which is often home to your email, social networking, and other online accounts — including, potentially, your files and financial information. Don't take phone security lightly.

Lock Your Phone

The first line of defense for your Android phone is the screen lock. It can be simple, it can be complex, or it can be nonexistent. The choice is up to you.

Finding the screen locks

All screen locks on your phone are found in the Settings app, on the Choose Screen Lock screen. Heed these steps to visit that screen:

1. **Open the Settings app.**

2. **Choose Security.**

   This item may have another name, such as Lock Screen and Security. If you see both Security and Lock Screen items, choose Lock Screen.

On some Samsung phones, you choose the Lock Screen item on the Device tab in the Settings app.

3. **Choose Screen Lock.**

   The item might also be titled Screen Lock Type, Set Up Screen Lock, or Change Screen Lock.

4. **Work any secure screen lock to continue.**

   Eventually, you see the Choose Screen Lock screen, which might instead be called Select Screen Lock.

The Choose Screen Lock screen lists several types of screen locks. Some are unique to your phone, and others are common Android screen locks, which include

- **Swipe:** Unlock the phone by swiping your finger across the touchscreen. This item might also be titled Slide.

- **Pattern:** Trace a pattern on the touchscreen to unlock the phone.

- **PIN:** Unlock the phone by typing a personal identification number (PIN).

- **Password:** Type a password to unlock the phone.

Other screen locks may be available, including Face unlock, fingerprint, signature, and even the None lock, which isn't a lock at all.

- The most secure lock types are the PIN and password. Either screen lock type is required if the phone has multiple users or is a kid's account or accesses a secure email server.

- The screen lock doesn't appear when you answer an incoming phone call. You're prompted, however, to unlock the phone if you want to use its features while you're on a call.

- I know of no recovery method available should you forget your phone's PIN or password screen locks. If you use either one, write it down in an inconspicuous spot, just in case.

Removing a screen lock

You don't remove the screen lock on your Android phone. Instead, you replace it. Specifically, to remove the pattern, PIN, or password screen lock, set the swipe lock. Follow the directions in the preceding section to get to the Choose Screen Lock screen.

- You may be prompted for confirmation if you're opting to reset a secure screen lock to one that's less secure.

- The phone prohibits you from removing a secure screen lock if the device is encrypted or accesses secure email or when other features are enabled.

Setting a PIN

The PIN lock is second only to the password lock as the most secure for your Android phone. To access the phone, you must type a PIN, or personal identification number. This type of screen lock is also employed as a backup for the less-secure screen unlocking methods, such as the pattern lock.

The *PIN lock* is a code between 4 and 16 numbers long. It contains only numbers, 0 through 9. To set the PIN lock for your Android phone, follow the directions in the earlier section "Finding the screen locks" to reach the Choose Screen Lock screen. Choose PIN from the list of locks.

Use the onscreen keypad to type your PIN once, and then tap the Continue button. Type the same PIN again to confirm that you know it. Tap OK. The next time you unlock the phone, you need to type the PIN to gain access.

Applying a password

The most secure way to lock the phone's screen is to apply a full-on password. Unlike a PIN, a *password* contains a combination of numbers, symbols, and uppercase and lowercase letters.

Set the password by choosing Password from the Choose Screen Lock screen; refer to the earlier section "Finding the screen locks" for information on getting to that screen. The password you create must be at least four characters long. Longer passwords are more secure.

You're prompted to type the password whenever you unlock your phone or whenever you try to change the screen lock. Tap the OK button to accept the password you've typed.

Creating an unlock pattern

Perhaps the most popular, and certainly the most unconventional, screen lock, is the pattern lock. The pattern must be traced on the touchscreen to unlock the phone. To create an unlock pattern, follow these steps:

1. **Summon the Choose Screen Lock screen.**

   Refer to the earlier section "Finding the screen locks."

2. **Choose Pattern.**

   If you haven't yet set a pattern, you may see the tutorial describing the process; tap the Next button to skip merrily through the dreary directions.

3. Trace an unlock pattern.

Use Figure 21-1 as your guide. You can trace over the dots in any order, but you can trace over a dot only once. The pattern must cover at least four dots.

I began the pattern here   Keep tracing

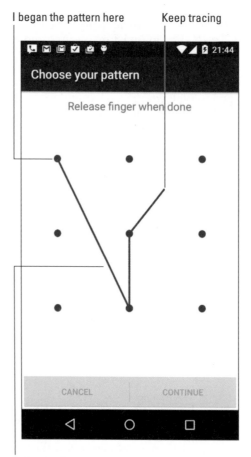

Pattern so far

Figure 21-1: Set the unlock pattern.

4. Tap the Continue button.

5. Redraw the pattern.

6. Tap the Confirm button.

You may be required to type a PIN or password as a backup to the pattern lock. If so, follow the onscreen directions to set that lock as well.

To ensure that the pattern appears on the Lock screen, place a check mark by the Make Pattern Visible option. For even more security, you can disable this option, but you must remember how — and where — the pattern goes.

Also: Clean the touchscreen! Smudge marks can betray your pattern.

Setting unusual screen locks

Your Android phone may offer other screen locks beyond the conventional locks described in this chapter. These include silly or fancy locks — perhaps not that secure but fun and different.

Among the more unusual screen lock types are face unlock, signature lock, and fingerprint lock. Choose these screen locks from the Choose Screen Lock screen. Work through the setup process. You may also need to set a PIN or password as a backup to the less-secure screen lock types.

- The face unlock works by using the phone's front camera. To unlock the device, you stare at the screen. So long as you haven't had any recent, major plastic surgery, the phone unlocks.

- The signature lock is unique to the Samsung Galaxy Note phone. Use the S-Pen to scribble your John Hancock on the touchscreen. The phone unlocks.

- Phones equipped with a fingerprint scanner allow you to swipe your favorite digit across the scanner gizmo to unlock the device. For example, the Samsung Galaxy S6 uses the Home button as the fingerprint scanner.

Other Phone Security

Beyond locking the screen, multiple other tools exist to help you thwart the Bad Guys and keep safe the information in your phone. Beyond that, you can even employ methods to locate a lost phone or display a phone that's been stolen.

Controlling Lock screen notifications

Your phone may allow for notifications to be displayed on the Lock screen, similar to how notifications are listed on the notifications drawer. You can set which notifications appear and how they're displayed, which helps add to the phone's security.

To configure Lock screen notifications, follow these steps:

1. **Open the Settings app.**

2. **Choose Sound & Notification.**

   This item might be titled Sounds and Notifications.

3. **Choose When Device Is Locked.**

   Another title for this setting is Notifications on Lock Screen.

4. **Select a Lock screen notification level.**

   Up to three settings are available:

   - Show All Notification Content

   - Hide Sensitive Notification Content

   - Don't Show Notifications at All

   The names of these settings may be subtly different on your phone.

5. **Choose a notification level.**

If your phone features a secure screen lock, the Hide Sensitive Notification Content option appears (refer to Step 4). When chosen, the Lock screen displays notices, not message previews or content.

To make the Lock screen notifications useful, double-tap one. The app that generated the notification starts, and you can see more details. Of course, you have to first work the screen lock.

Adding owner info text

Suppose that you lose your Android phone. Wouldn't it be nice if a good Samaritan found it? What would be even more helpful is information on the Lock screen to help that kind person find you and return your gizmo. That information is called the *owner info text*.

To add owner info text to your Android phone's Lock screen, follow these steps:

1. **Open the Settings app.**

2. **Choose the Security or Lock Screen category.**

   On some Samsung phones, the Lock Screen category is found on the Device tab.

3. **Choose Owner Info or Owner Information.**

4. **Ensure that there's a check mark by the option Show Owner Info on Lock Screen.**

5. **Type text in the box.**

   You can type more than one line of text, though the information is displayed on the Lock screen as a single line.

6. **Tap the OK button.**

   Or tap the Back navigation icon to return to the previous screen.

Whatever text you type in the box appears on the Lock screen. Therefore, I recommend typing something useful: your name, address, another phone number, an email address, or similar vital information. This way, should you lose your phone and an honest person find it, he can get it back to you.

The owner info may not show up when None is selected as a screen lock.

Finding a lost phone

Someday, you may lose your beloved Android phone. It might be for a few panic-filled seconds, or it might be for forever. The hardware solution is to weld a heavy object to the phone, such as an anvil. Alas, that strategy kind of defeats the entire mobile/wireless paradigm.

To quickly locate your Android phone, follow these steps while using a computer:

1. **Open a web browser, such as Google Chrome.**

2. **Visit the main Google search page:** www.google.com.

3. **Type** find my phone **and press the Enter key.**

4. **If prompted, sign in to your Google (Gmail) account.**

   Your phone's location appears on the screen.

Another solution is to use a cell phone locator service.

Cell phone locator services employ apps that use the phone's mobile data signal as well as its GPS to help locate the missing gizmo. These types of apps are available at the Google Play Store. One that I've tried and recommend is Lookout Mobile Security.

Lookout features two different apps. One is free, which you can try to see whether you like it. The paid app offers more features and better locating services. As with similar apps, you must register at a website to help you locate your phone, should it go wandering.

If you use a Samsung phone, check the Settings app, the Lock Screen and Security item, for the Find My Mobile feature. You must have a Samsung account to enable this feature, but it's a better choice than a third-party app.

Encrypting the phone

When the information on your beloved phone must be really, really secure, you can take the drastic step of encrypting its storage. I heard a rumor that the secret recipe for Coca-Cola is held in such a manner.

Before considering this option, be aware that it's not currently possible to remove encryption. After encryption is applied, it's stuck forever like a regrettable, drunken tattoo.

Start the encryption process by applying a secure screen lock, such as the PIN or password.

Second, ensure that the phone is either plugged in or fully charged. Encryption takes a while — up to several hours if your phone's storage is rather full — and you don't want your gizmo pooping out before the process is complete.

Third, follow these steps to encrypt the phone's internal storage:

1. **Open the Settings app and choose Security.**

   On some Samsung phones, choose Lock Screen and Security and then Other Security Settings.

2. **Choose Encrypt Phone.**

   If you haven't followed my advice in this this section, you need to charge the phone or set a password or both.

3. **Tap the Encrypt Phone button.**

4. **Wait.**

Once encrypted, only by unlocking the phone can anyone access its storage. Further, you need to work the screen lock every time you turn on the phone.

My advice is not to encrypt your phone but to instead use a strong password.

Performing a factory data reset

The most secure thing you can do with the information on your Android phone is to erase it all. The procedure, known as a *factory data reset,* effectively restores the device to its original state, fresh out of the box.

A factory data reset is a drastic thing. It removes not only all information from storage but also erases all your accounts. Don't take this step lightly! In fact, if you're using this procedure to cure an ill, I recommend first getting support. See Chapter 23.

When you're ready to erase all the phone's data, follow these steps:

1. **Start the Settings app and choose Backup & Reset.**

   On Samsung phones, tap the General tab to locate the Backup and Reset item.

2. **Choose Factory Data Reset.**

3. **Tap the Reset Phone button.**

4. **If prompted, work the screen lock.**

   This level of security prevents others from idly messing with your beloved phone.

5. **Tap the Erase Everything or Delete All button to confirm.**

   All the information you've set or stored on the phone is purged, including all your accounts, any apps you've downloaded, music — everything.

Practical instances when this action is necessary include selling your phone, giving it to someone else to use, and exchanging it for a new phone.

On the Road Again

In This Chapter

▶ Understanding roaming

▶ Placing international calls

▶ Entering Airplane mode

▶ Using the phone in the air

▶ Contacting friends abroad

▶ Going overseas with your phone

*Y*ou're in a land far, far away. The sun shines warmly upon your face. A gentle breeze wafts over crashing waves. You wiggle your toes in the soft, grainy sand. And the number-one thought on your mind is, "Can my Android phone get a signal?"

The vernacular term is *bars,* as in the signal bars you see when the phone is actively communicating with the mobile data network. That network pretty much covers the globe these days. That means from Cancun to the Gobi Desert, you'll probably be able to use your phone. How that happens and how to make it happen less expensively are the topics of this chapter.

20:43

SAT, DECEMBER 28

Charging, 51%

Where the Phone Roams

The word *roam* takes on an entirely new meaning when applied to a cell phone. It means that your phone receives a cell signal whenever you're outside the phone carrier's operating area. In that case, your phone is *roaming.*

Roaming sounds handy, but there's a catch: It almost always involves a surcharge for using another cellular service — an *unpleasant* surcharge.

Relax: Your Android phone alerts you whenever it's roaming. The Roaming icon appears at the top of the screen, in the status area, whenever you're outside your cellular provider's signal area. The icon differs from phone to phone, but generally the letter *R* figures in it somewhere, similar to what's shown in the margin.

There's little you can do to avoid incurring roaming surcharges when making or receiving phone calls. Well, yes: You can wait until you're back in an area serviced by your primary cellular provider. You can, however, altogether avoid using the other mobile data network while roaming. Follow these steps:

1. **Open the Settings app.**

2. **Under the Wireless & Networks heading, tap the More item and choose Cellular Networks.**

   If you see a Mobile Networks item in the Wireless and Networks area, choose it instead. Choose the item Global Data Roaming Access.

3. **Ensure that the Data Roaming setting is disabled or denied.**

Your phone can still access the Internet over the Wi-Fi connection when it roams. Setting up a Wi-Fi connection doesn't affect the mobile data network connection, because the phone prefers to use Wi-Fi. See Chapter 17 for more information about Wi-Fi.

Another network service you might want to disable while roaming has to do with multimedia, or *MMS,* text messages. To avoid surcharges from another cellular network for downloading an MMS message, follow these steps:

1. **Open the phone's text messaging app, such as Google Hangouts.**

   See Chapter 8 for details on using that app.

2. **Tap the Side Menu icon to view the app's navigation drawer and choose Settings.**

   The Settings action might also be accessed by tapping the Action Overflow icon.

3. **Choose SMS.**

4. **Ensure that the Roaming Auto-Retrieve option is disabled.**

The lock screen may also announce that the phone is roaming. You might see the name of the other cellular network displayed. The text *Emergency Calls Only* might also appear.

 If you're greatly concerned about roaming while overseas, simply place the phone into Airplane mode, as discussed elsewhere in this chapter.

International Calling

A phone is a bell that anyone in the world can ring. To prove it, all you need is the phone number of anyone in the world. Dial that number by using your phone and, as long as you both speak the same language, you're talking!

To make an international call with your Android phone, you merely need to know the foreign phone number. The number includes the international country-code prefix, followed by the number. For example:

01-234-56-789

Before dialing the international country-code prefix (01 in this example), you must first type a plus sign (+). The + symbol is the *country exit code,* which must be dialed to flee the national phone system and access the international phone system. For example, to dial Finland on your phone, type **+358** and then the number in Finland. The +358 is the exit code (+) plus the international code for Finland (358).

To type the + character, press and hold down the 0 key on the Phone app's dialpad. Then type the country prefix and the phone number. Tap the Dial icon to place the call.

✔ Dialing internationally involves surcharges, unless your cell plan provides for international dialing.

✔ International calls fail for a number of reasons. One of the most common is that the recipient's phone service blocks incoming international calls.

✔ Another reason that international calls fail is the zero reason: Oftentimes, you must leave out any zero in the phone number that follows the country code. So, if the country code is 254 for Kenya and the phone number starts with 012, you dial +254 for Kenya and then 12 and the rest of the number. Omit the leading zero.

✔ Know which type of phone you're calling internationally — cell phone or landline. The reason is that an international call to a cell phone might involve a surcharge that doesn't apply to a landline.

✔ The + character isn't a number separator. When you see an international number listed as 011+20+xxxxxxx, do not insert the + character in the number. Instead, type **+20** and then the rest of the international phone number.

✔ Most cellular providers add a surcharge when sending a text message abroad. Contact your cellular provider to confirm the text message rates. Generally, you find two rates: one for sending and another for receiving text messages.

✔ If texting charges vex you, remember that email has no associated per-message charge. You also have alternative ways to chat, such as Google Hangouts and Skype; Skype can also be used to place cheap international calls. See Chapter 11.

✔ In most cases, dialing an international number involves a time zone difference. Before you dial, be aware of what time it is in the country or location you're calling. The Clock app can handle that job for you: Summon a clock for the location you're calling and place it on the Clock app's screen.

You Can Take It with You

Don't let them fool you: If you're asked how far you can throw your Android phone, just don't do it. It's a trick foisted by the jealous owners of I-fruit phones. The truth is that you can take your Android phone with you anywhere you like. How it functions may change depending on your environment, and you can do a few things to prepare before you go, which are all covered in this section.

Preparing to leave

Unless you're being unexpectedly abducted, you should prepare several things before leaving on a trip with your phone.

First and most important, of course, is to charge the thing. You probably charge your phone overnight anyway, but if you're leaving in the middle of the day, plug in your phone to ensure that the charge will last the journey.

Second, consider loading up on some reading material, music, and a few new apps before you go. Nothing beats getting eBooks for the road. You can also preload or save some web pages for offline viewing, as discussed in Chapter 10. Visit the Play Store to shop for movies, music, and books before you go. Perhaps even load up a new game that you can learn to play.

✔ If you plan to read books, listen to music, or watch a video while on the road, consider downloading that media to your phone before you leave. See Chapter 16 for information.

✔ Most major airlines offer travel apps. These apps generate notifications for your schedule and provide timely gate changes or flight delays; plus, you can use the phone as your e-ticket. Search the Play Store to see whether your preferred airline offers an app.

Arriving at the airport

I'm not a frequent flier, but I am a nerd. The most amount of junk I've carried with me on a flight is two laptop computers and three cell phones. I know that's not a record, but it's enough to warrant the following list of travel tips, all of which apply to taking your phone with you on an extended journey:

✔ Bring the phone's AC adapter and USB cable with you. Put them in your carry-on luggage.

✔ Nearly all major airports feature USB chargers, so you can charge the phone in an airport, if you need to. Even though you need only the cable to charge, bring along the AC adapter anyway.

✔ At the security checkpoint, place your phone in a bin by itself or with other electronics (tablets, laptops). Even if you've been preapproved, you must remove the phone from your pocket.

✔ Open the phone's web browser to use the airport's Wi-Fi service. Most airports don't charge for the service, although you may have to agree to terms by visiting the airport's website and watching an advertisement or agreeing to the service terms.

Flying with your phone

After the crew closes the airplane's door, you must turn off your phone or place it into Airplane mode. You can still use the phone, just not to place phone calls or text messages.

Follow these steps to place your phone into Airplane mode:

1. **Open the Settings app.**

2. **Choose the More item in the Wireless & Networks section.**

   If you find an Airplane Mode item in the Wireless and Networks section, choose that item instead.

3. **Enable Airplane mode.**

   Slide the master control to the On position, or place a check mark by the Airplane Mode item.

While the phone is in Airplane mode, a special icon appears in the status area, similar to the one shown in the margin. You might also see the text *No Service* on the phone's Lock screen.

✔ You can quickly enter Airplane Mode by choosing its Quick Setting: Swipe down the notifications drawer and choose Airplane Mode. See Chapter 3 for more details on accessing your phone's Quick Settings.

✔ Some phones feature an Airplane Mode action on the Device Options menu: Press and hold the Power Lock key as though you're turning off the phone. Choose Airplane Mode.

✔ Some Android phones feature Sleep mode, in which case you can place your phone into Sleep mode for the duration of a flight. The phone wakes up faster from Sleep mode than it does when you turn it on.

✔ You can still use Wi-Fi while aloft, as long as that service is available and you're willing to pay for it. You must activate Wi-Fi after enabling Airplane mode. See Chapter 17 for information on turning on the phone's Wi-Fi radio.

✔ Airplane mode disables all of the phone's wireless radios: the cellular radio, Wi-Fi, Bluetooth, NFC, and GPS. You can, however, reenable Wi-Fi, although using the other radios is forbidden while flying.

Getting to your destination

After you arrive at your destination, the phone may update the date and time according to your new location. One additional step you may want to take is to set the phone's time zone. By doing so, you ensure that your schedule adapts properly to your new location.

To change the phone's time zone, follow these steps:

1. **Open the Settings app.**

2. **Choose Date and Time.**

   On some Samsung phones, tap the General tab to locate the Date and Time item.

3. **Ensure that the Automatic Time Zone setting is enabled.**

If the phone's time isn't updated, continue by choosing the Select Time Zone item. Pluck the current time zone from the list.

If you've set appointments for your new location, visit the Calendar app to ensure that their start and end times have been properly adjusted. If you're prompted to update appointment times based on the new zone, do so.

When you're done traveling, or when you change time zones again, ensure that the phone is updated as well. When the Automatic Time Zone setting isn't available, follow the steps in this section to reset the phone's time zone.

The Android Phone Goes Abroad

Yes, your Android phone works overseas. The world uses similar types of cellular networks, so the odds are good that your phone will work immediately in whichever county you find yourself. Your concerns should be overseas roaming, charging the phone in a foreign country, and how to get on the Internet.

Using your Android phone overseas

The easiest way to use a cell phone abroad is to rent or buy one in the country where you plan to stay. I'm serious: Often, international roaming charges are so high that it's cheaper to simply buy a temporary cell phone wherever you go, especially if you plan to stay there for a while.

When you opt to use your own phone rather than buy a local phone, things should run smoothly — if a compatible cellular service is in your location. Not every Android phone uses the same network and, of course, not every foreign country uses the same cellular network. Things must match before the phone can work. Plus, you may have to deal with foreign carrier roaming charges.

The key to determining whether your phone is usable in a foreign country is to turn it on. The name of that country's compatible cellular service should show up on the phone's Lock screen. So where your phone once said *Verizon Wireless,* it may say *Wambooli Telcom* when you're overseas.

- ✔ You receive calls on your cell phone internationally as long as the phone can access the network. Your friends need only dial your cell phone number as they normally do; the phone system automatically forwards your calls to wherever you are in the world.

- ✔ The person calling you pays nothing extra when you're off romping the globe with your Android phone. Nope — *you* pay extra for the call.

- ✔ While you're abroad, you need to dial internationally. When calling home (for example, the United States), you need to use a 10-digit number (phone number plus area code). You may also be required to type the country exit code when you dial. See the earlier section "International Calling."

- ✔ When in doubt, contact your cellular provider for tips and other information specific to whatever country you're visiting.

- ✔ Be sure to inquire about texting and cellular data (Internet) rates while you're abroad.

- ✔ Also see the section "Where the Phone Roams," earlier in this chapter. Roaming charges definitely apply overseas.

Charging the phone in another country

You can easily attach a foreign AC power adapter to your phone's AC power plug. You don't need a voltage converter; just an adapter. After the adapter is attached, plug your phone into those weirdo overseas power sockets without the risk of blowing up anything.

Accessing Wi-Fi in foreign lands

Using an Android phone over a Wi-Fi network abroad incurs no extra fees (unless data roaming is on, as discussed earlier in this chapter). The same protocols and standards are used everywhere, so if the phone can access Wi-Fi at your local Starbucks, it can access Wi-Fi at the Malted Yak Blood Café in Wamboolistan. As long as Wi-Fi is available, your Android phone can use it.

✔ Internet cafés are more popular overseas than in the United States. They are the best locations for connecting your phone and catching up on life back home.

✔ Many overseas hotels offer free Wi-Fi service, although the signal may not reach into every room. Don't be surprised if you can use the Wi-Fi network only while you're in the lobby.

✔ Use your Android phone to make phone calls overseas by getting some Skype Credit. Skype's international rates are quite reasonable. The calls are made over the Internet, so when the phone has Wi-Fi access, you're good to go. See Chapter 11 for more information on making Skype calls.

Maintenance, Troubleshooting, and Help

Maintenance is that thing you were supposed to remember to do but you didn't do, and that's why you need help and troubleshooting advice. Don't blame yourself; no one likes to do maintenance. Okay, well, I like maintaining my stuff. I even change the belt on my vacuum cleaner every six months. Did you know that the vacuum cleaner manual tells you to do so? Probably not. I read that in *Vacuum Cleaners For Dummies*. This book is *Android Phones For Dummies*, which is why it contains topics on maintenance, troubleshooting, and help for Android phones, not for vacuum cleaners.

Regular Phone Maintenance

Relax. Unlike draining the lawnmower's oil once a year, regular maintenance of an Android phone doesn't require a drip pan or a permit from the EPA. In fact, an Android phone requires only two basic regular maintenance tasks: cleaning and backing up.

Keeping it clean

You probably already keep your phone clean. I must use my sleeve to wipe the touchscreen at least a dozen times a day. Of course, better than your sleeve is something called a *microfiber cloth*. This item can be found at any computer or office-supply store.

✔ Never use ammonia or alcohol to clean the touchscreen. These substances damage the phone. If you must use a cleaning solution, select something specifically designed for touchscreens.

✔ If the screen continually gets dirty, consider adding a *screen protector*. This specially designed cover prevents the screen from getting scratched or dirty while still allowing you to interact with the touchscreen. Ensure that the screen protector you obtain is intended for use with your specific phone.

✔ You can also find customized cell phone cases, belt clips, and protectors, which can help keep the phone looking spiffy. Be aware that these items are mostly for decorative or fashion purposes and don't even prevent serious damage should you drop the phone.

Backing up your phone

For a majority of the information on your phone, backup is automatic. Your Google account takes care of the Gmail, calendar, contacts — even your apps. If you don't download music, eBooks, or movies, nothing ever needs to be backed up, at least not manually.

To confirm that your phone's information is being backed up, heed these steps:

1. **Open the Settings app and choose Accounts.**

   On some phones, the accounts are listed right on the main Settings app screen.

2. **Choose your Google account.**

3. **Tap your Gmail address.**

4. **Ensure that each item in the list is enabled.**

   These are the items that synchronize between the phone and your Google account on the Internet.

   You're not done yet!

5. **Tap the Back navigation until you're returned to the main Settings app screen.**

6. **Choose Backup & Reset.**

7. **Ensure that the item Back Up My Data is enabled.**

Beyond your Google account, which is automatically backed up, the rest of the information can be manually backed up. You can copy files from the phone's internal storage to the cloud or your computer as a form of backup. See Chapter 18 for information on coordinating files between your phone and a computer.

Updating the system

Every so often, a new version of your phone's operating system becomes available. It's an *Android update* because *Android* is the name of the phone's operating system, not because your phone thinks that it's a type of robot.

Whenever an update occurs, you see an alert or a message indicating that a system upgrade is available. The message may be as subtle as a notification icon, or it might be a card onscreen informing you that an update is necessary. My advice: Install the update and get it over with. Don't dally.

✔ If possible, connect the phone to a power source during a software update. You don't want the battery to die in the middle of the operation.

✔ You can manually check for updates: In the Settings app, choose About Phone. Choose System Updates or Software Update. When the system is up-to-date, the screen tells you so. Otherwise, you find directions for updating the Android operating system.

✔ Tapping the Check Now button isn't magic. When an update is available, the phone lets you know.

✔ Non-Android system updates might also be issued. For example, the phone's manufacturer may send out an update to the device's guts. This type of update is often called a *firmware* update. As with Android updates, my advice is to accept all firmware updates.

Battery Care and Feeding

Perhaps the most important item you can monitor and maintain on your Android phone is its battery. The battery supplies the necessary electrical juice by which the phone operates. Without battery power, your phone is about as useful as a tin-can-and-a-string for communications. Keep an eye on the battery.

Monitoring the battery

Your phone displays its current battery status at the top of the screen, in the status area, next to the time. The icons used are similar to those shown in Figure 23-1.

Battery is fully charged and happy.

Battery is being used, starting to drain.

Battery getting low, you should charge!

Battery frighteningly low, stop using and charge at once!

Battery is charging.

Figure 23-1: Battery status icons.

You might also see the icon for a dead or missing battery, but for some reason I can't get my phone to turn on and display that icon.

- Heed those low-battery warnings! The phone generates a notification whenever the battery power gets low. The phone generates another notification whenever the battery gets *very* low.

- When the battery is too low, the phone shuts itself off.

- The best way to deal with a low battery is to connect the phone to a power source: Either plug the phone into a wall socket, or connect the phone to a computer by using a USB cable. The phone charges itself immediately; plus, you can use the phone while it's charging.

- The phone charges more efficiently when it's plugged into a wall socket rather than a computer.

- You don't have to fully charge the phone to use it. If you have 20 minutes to charge and the power level returns to only 70 percent, that's great. Well, it's not great, but it's far better than a 20 percent battery level.

> ✔ Battery percentage values are best-guess estimates. Just because you talked for two hours and the battery shows 50 percent doesn't mean that you're guaranteed two more hours of talking. Odds are good that you have much less than two hours. In fact, as the percentage value gets low, the battery appears to drain faster.

Determining what is drawing power

The phone's Battery screen informs of you battery usage over time, as well as which of your apps have been consuming power. Figure 23-2 illustrates such a screen.

To view the Battery screen on your phone, open the Settings app and choose the Battery item. On some Samsung phones, tap the General tab in the Settings app to locate the Battery item.

Tap an item in the list to view its details. For some items, the details screen lets you adjust power settings.

The number and variety of items listed on the battery use screen depend on what you've been doing with your phone between charges and how many different apps you've been using. Don't be surprised if an item (such as the Play Books app) doesn't show up in the list. Not every app uses a lot of battery power.

Extending battery life

Here's a smattering of things you can do to help prolong battery life for your Android phone:

Dim the screen. The display is capable of drawing down quite a lot of battery power. Although a dim screen can be more difficult to see, especially outdoors, it definitely saves on battery life. See Chapter 20 for information on adjusting the screen brightness.

Disable vibration options. The phone's vibration is caused by a teensy motor. Although you don't see much battery savings by disabling the vibration options, it's better than no savings. To turn off vibration, follow these steps:

1. **Open the Settings app and choose Sound & Notification.**

2. **Disable one or all of the vibration settings.**

   Most phones lack a single setting for vibration. Instead, you find settings such as Vibrate While Ringing, Vibration Intensity, and Vibrate on Touch.

Current battery charge and state

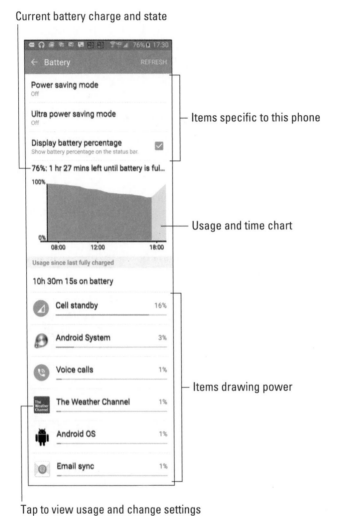

Items specific to this phone

Usage and time chart

Items drawing power

Tap to view usage and change settings

Figure 23-2: Battery usage info.

Deactivate Bluetooth. When you're not using Bluetooth, turn it off. See Chapter 17 for information on Bluetooth.

Manage battery performance. Many Android phones have battery-saving software built in. You can access the software from a special app or from the Battery item in the Settings app. Similar to your computer, battery-performance management involves turning phone features on or off during certain times of the day. Third-party battery management apps are also available at the Google Play Store. See Chapter 16.

Help and Troubleshooting

Getting help isn't as bad as it was in the old days. Back then, you could try two sources for technological help: the atrocious manual that came with your electronic device or a phone call to the guy who wrote the atrocious manual. It was unpleasant. Today, the situation is better. You have many resources for solving issues with your gizmos, including your Android phone.

Fixing random and annoying problems

Aren't all problems annoying? A welcome problem doesn't exist, unless the problem is welcome because it diverts attention from another, preexisting problem. And random problems? If problems were predictable, they would serve in office.

General trouble

For just about any problem or minor quirk, consider restarting the phone: Turn off the phone, and then turn it on again. This procedure will likely fix a majority of the annoying and quirky problems you encounter when using an Android phone.

Some Android phones feature the Restart action on the Phone Options menu: Press and hold down the Power/Lock key to see this menu. If a Restart command is present, use it to restart the phone and fix (you hope) whatever has gone awry.

Connection woes

Sometimes the data connection drops but the phone connection stays active. Check the status bar. If you see bars, you have a phone signal. When you don't see the 4G, 3G, 1X, or Wi-Fi icon, the phone has no Internet access.

Do expect the mobile data signal to change. You may observe the status icon change from 4G LTE to 3G to even the dreaded 1X or — worse — nothing, depending on the strength and availability of the mobile data network. Wait, and it should come back around. If not, the mobile data network might be down, or you may simply be in an area with lousy service. Consider changing your location.

For Wi-Fi connections, ensure that Wi-Fi is set up properly and working. This process usually involves pestering the person who configured the Wi-Fi router or, in a coffee shop, bothering the cheerful person with the tattoos and piercings who serves you coffee.

Be aware that some Wi-Fi networks have a "lease time" after which the phone is disconnected. If so, follow the directions in Chapter 17 for turning off Wi-Fi and then turn it on again. That often solves the issue.

Another problem could be that the Wi-Fi router doesn't recognize your phone. It may be an older router that can't handle an abundance of wireless devices. If this situation describes your home or office network, and the Wi-Fi router is over five years old, consider getting a newer router.

Music is playing and you want it to stop

It's awesome that the phone continues to play music while you do other things. Getting the music to stop quickly, however, requires some skill. You can access the play controls for the Play Music app from a number of locations. They're found on the Lock screen, for example. You can also find them on the notifications drawer.

An app has run amok

Sometimes, apps that misbehave let you know. You see a warning on the screen announcing the app's stubborn disposition. When that happens, tap the Force Quit button to shut down the app. Then say "Whew!"

To manually shut down an app, refer to Chapter 19.

You've reached your wit's end

When all else fails, you can do the drastic thing and perform a factory data reset on your phone. Before committing to this step, I recommend that you contact support as described in the next section.

Refer to Chapter 21 for details on the factory data reset.

Getting help and support

Never discount two important sources of support for your Android phone: the cellular provider and the phone's manufacturer. Between the two, I recommend contacting your cellular provider first, no matter what the problem.

The Help app

Some phones come supplied with the Help app. The app may be called Help or Help Center or something similar, and it may not be the kind of avuncular, well-written help you get from this book, but it's better than nothing.

- ✔ A Guided Tour or Tutorial app may also be available, which helps you understand how to work some of the phone's interesting features.

- ✔ Also look for a Help document or phone manual eBook in the Play Books app.

 ✔ The Settings app features the Search icon, which helps you locate specific settings without knowing exactly under which category the item might be found.

Cellular support

Contact information for both the cellular provider and phone manufacturer is found in the material you received with the phone. In Chapter 1, I recommend that you save those random pieces of paper. You obviously have read that chapter and followed my advice, so you can easily find that information.

Okay, so you don't want to go find the box, or you didn't heed my admonition and you threw out the box. Table 23-1 lists contact information on U.S. cellular providers. The From Cell column lists the number you can call by using your Android phone; otherwise, you can use the toll-free number from any phone.

Table 23-1		U.S. Cellular Providers	
Provider	*From Cell*	*Toll-free*	*Website*
AT&T	611	800-331-0500	www.att.com/esupport
Sprint Nextel	*2	800-211-4727	mysprint.sprint.com
T-Mobile	611	800-866-2453	www.t-mobile.com/Contact.aspx
Verizon	611	800-922-0204	http://support.vzw.com/clc

Your cellular provider's Help phone number might already be found in the phone's address book app. Look for it there if it's not listed in the table.

Manufacturer support

The second source of support is the phone's manufacturer, such as Samsung or LG. Information about support can be found in those random papers and pamphlets included in the phone's box. If not, refer to Table 23-2 for contact information for various Android handset manufacturers.

Table 23-2	Android Phone Manufacturers
Manufacturer	*Website*
HTC	www.htc.com/us/support
LG	www.lg.com/us/support
Motorola	www.motorola.com
Samsung	www.samsung.com/us/mobile/cell-phones

App support

For app issues, contact the developer. Unlike finding developer support for computer software, accessing an app developer is cinchy. Follow these steps:

1. **Open the Play Store app.**

2. **Tap the Side Menu icon to display the navigation drawer.**

3. **Choose My Apps.**

4. **Tap the entry for the specific app, the one that's bothering you.**

5. **Choose the Send Email item.**

    Swipe the screen from button to top to scroll down the app's info screen and locate the Send Email item.

Contacting the developer is no guarantee that you'll get a response.

Play Store support

For issues with the Play Store itself, contact Google at `support.google.com/googleplay`.

Android Phone Q&A

I love Q&A! That's because not only is it an effective way to express certain problems and solutions but some of the questions might also cover topics I've been wanting to write about.

"The touchscreen doesn't work!"

A touchscreen requires a human finger for proper interaction. The phone interprets complicated electromagnetic physics between the human finger and the phone to determine where the touchscreen is being touched. Bottom line: You can't wear gloves and use the touchscreen unless they are specific touchscreen gloves.

The touchscreen might also fail when the battery charge is low or when the phone has been physically damaged.

"The screen is too dark!"

Android phones feature a teensy light sensor on the front. If the phone's Adaptive Brightness feature is active, the sensor adjusts the touchscreen's brightness based on the amount of ambient light at your location. If the sensor is covered, the screen can get very, very dark.

Ensure that you aren't unintentionally blocking the light sensor. Avoid buying a case or screen protector that obscures the sensor.

Also see Chapter 20 for information on setting the screen brightness.

"The battery doesn't charge!"

Start from the source: Is the wall socket providing power? Is the cord plugged in? The cable may be damaged, so try another cable.

When charging from a USB port on a computer, ensure that the computer is turned on. Computers provide no USB power when they're turned off. The phone may charge only when connected to a powered USB hub, such as those found directly on the computer console.

"The phone gets so hot that it turns itself off!"

Yikes! An overheating phone can be a nasty problem. Judge how hot the phone is by seeing whether you can hold it in your hand: When the phone is too hot to hold, it's too hot. If you're using the phone to keep your coffee warm, the phone is too hot.

Turn off the phone and let it cool.

If the overheating problem continues, have the phone looked at for potential repair.

"The phone won't do Landscape mode!"

Just because an app doesn't enter Landscape mode doesn't mean that it *can* enter Landscape mode. Not every app takes advantage of the phone's capability to orient itself in Landscape mode. On certain Android phones, the Home screen doesn't "do landscape."

One app that definitely does Landscape mode is the web browser, which is described in Chapter 10.

Android phones have a Quick Setting you can check to confirm that screen rotation is enabled. See Chapter 3 for information on Quick Settings.

Part V
The Part of Tens

Enjoy a Part of Tens list online at
www.dummies.com/extras/androidphones.

In this part . . .

- Glean ten tips, tricks, and shortcuts.
- Remember ten important things.

Ten Tips, Tricks, and Shortcuts

A tip is a small suggestion, something spoken from experience or insight and something you may not have thought of yourself. A *trick* is something that's good to know — something surprising. And a *shortcut* is convenient and quick, like using duct tape as a parenting tool.

Although I'd like to think that everything I mention in this book is a tip, trick, or shortcut for using your Android phone, I can offer even more information. This chapter provides ten tips, tricks, and shortcuts to help you get the most from your phone.

Quickly Switch Apps

Android apps don't quit. Sure, some of them have a Quit command or Sign Out command, but most apps lurk inside the phone's memory while you do other things. The Android operating system may eventually kill off a stale app. Before that happens, you can deftly and quickly switch between all running apps.

The key to making the switch is to use the Recent navigation icon, found at the bottom of the touchscreen. Figure 24-1 illustrates two popular representations of the Recent navigation icon.

After tapping the Recent navigation icon, choose an app from the list. Swipe the list up or down to peruse what's available. To dismiss the list, tap the Back or Home navigation icon.

Android Lollipop   **Android Kitkat**

Figure 24-1: Incarnations of the Recent navigation icon.

✔ On older phones that lack the Recent icon, long-press the Home navigation icon.

✔ To remove an app from the list of recent apps, swipe it left or right. This is almost the same thing as quitting an app.

✔ Some phones may feature a task manager, in the form of an app or, often, an icon at the bottom of the Recent Apps list.

✔ The difference between a task manager and the Recent Apps list is that the task manager lets you kill off running apps. This brutal step isn't necessary, but some people take great satisfaction in using the feature.

✔ The list of recent apps is called the Overview, although everyone I know calls it the list of recent apps.

Install Apps from a Computer

You don't need to use the Play Store app on your phone to install apps. Using a computer, you can visit the Google Play website and browse for apps and then have them installed remotely. It's kind of cool yet kind of scary at the same time. Here's how it works:

1. **Use a computer's web browser to visit the Google Play store on the Internet:** `play.google.com/store`.

   Bookmark that site!

2. **If necessary, click the Sign In button to log in to your Google account.**

   Use the same Google account you used when setting up your Android phone. You need to have access to that account so that Google can remotely update your various Android devices.

3. **Browse for something.**

   You can hunt down apps, books, music — the whole gamut. It works just like browsing the Play Store on your phone.

4. **Click the Install button.**

   The button is labeled Buy for paid apps.

5. **Choose your Android phone from the Choose a Device menu.**

   If your phone isn't listed, the app isn't compatible. That happens. Also, the phone may be listed using its technical name, not the brand name you're used to.

6. **For a free app, click the Install button. For a paid app, click the Continue button, choose your payment source, and then click the Buy button.**

   Installation proceeds.

As if by magic, the app is installed on your Android phone — even though you used a computer to do it. Heck, the phone need not even be within sight of you, and the app still installs remotely.

The Camera App's Panorama Mode

Most variations of the Camera app sport a panoramic shooting mode. The *panorama* is a wide shot — it works by panning the phone across a scene. The Camera app then stitches together several images to build a panoramic image.

To shoot a panoramic shot, follow these steps in the Camera app:

1. **Choose the Camera app's Panorama mode.**

   For the Google Camera app, swipe your finger inward from the left side of the screen. Tap the Panorama icon, shown in the margin, to switch the app into Panorama mode.

2. **Hold the phone steady.**

3. **Tap the Shutter icon.**

4. **Pivot in one direction, following the onscreen animation.**

   Watch as the image is rendered and saved.

Panoramas work best for vistas, wide shoots, or perhaps for family gatherings where not everyone likes everyone else.

Avoid Data Surcharges

An important issue for everyone using an Android phone is whether you're about to burst through your monthly data quota. Mobile data surcharges can pinch the wallet, but your Android phone has a handy tool to help you avoid data overages. It's the Data Usage screen, shown in Figure 24-2.

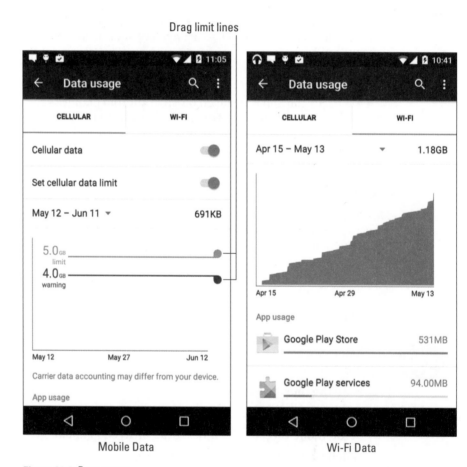

Figure 24-2: Data usage.

To access the Data Usage screen, follow these steps:

1. **Open the Settings app.**

2. **Choose Data Usage.**

   On some Samsung phones, you find the Data Usage item by choosing the Connections tab at the top of the Settings app screen.

The Data Usage screen sports two tabs: one for Cellular (mobile data) access and another for Wi-Fi. Each screen sports a line charge that graphically illustrates data usage over time. As you can see from Figure 24-2, my phone uses far more Wi-Fi than mobile data, which is fine by my wallet.

The red and black lines on the Cellular tab can help remind you of how quickly you're filling the phone's mobile data quota. A warning message appears when data usage crosses the black line. When the red line is reached, mobile data usage stops.

To set a warning (black line) limit, drag the tab on the right side of the screen up or down.

To set the stop (red line) limit, activate the Set Cellular Data Limit option; slide the master control icon to the On position. Then adjust the red line up or down.

To review access for a specific app, swipe the screen bottom-to-top to view the App Usage area. Only apps that access the network appear. Tap an app to view detailed data usage information. If you notice that the app is using more data than it should, tap the App Settings button. You may be able to adjust the app's Internet access, which also helps avoid data surcharges.

Make the Phone Dream

Does your phone lock, or does it fall asleep? I prefer to think that the phone sleeps. That begs the question of whether or not it dreams.

Of course it does! You can even see the dreams, if you activate the Daydream feature — and if you keep the phone connected to a power source or in a docking station. Heed these steps:

1. **Start the Settings app.**

2. **Choose Display and then Daydream.**

   The Display item is found on the Device tab on certain Samsung phones.

3. **Ensure that the Daydream master control is in the On position.**

4. **Choose which type of daydream you want displayed.**

    The Clock is a popular item, though I'm fond of Colors.

    Some daydreams feature a Settings icon, which can be used to customize how the daydream appears.

5. **Tap the Action Overflow and choose When to Daydream.**

6. **Choose the Either option.**

The daydreaming begins when the screen would normally time-out and lock, although the phone has to be receiving power. So if you've set the phone to lock after 1 minute of inactivity, it daydreams instead — provided that it's plugged in or docked.

✔ To disrupt the dream, swipe the screen.

✔ The phone doesn't lock when it daydreams. To lock the phone, press the Power/Lock key.

Charge the Battery Without Wires

When you want to go truly wireless (I mean, forgo all wires — period), take advantage of wireless charging. This can be a phone feature, or you can upgrade your phone by adding a wireless-charging battery cover.

After the phone is equipped for wireless charging, you simply set it on the charging pad or into the charging cradle. The battery charges. Voilà! No more wires.

Add Spice to Dictation

I feel that too few people use dictation, despite how handy it can be — especially for text messaging. Anyway, if you've used dictation, you might have noticed that it occasionally censors some of the words you utter. Perhaps you're the kind of person who doesn't put up with that kind of s***.

Relax. You can lift the vocal censorship ban by following these steps:

1. **Start the Settings app.**

2. **Choose Language & Input.**

    On some Samsung phones you'll find the Language and Input item on either the Controls tab or the General tab in the Settings app.

3. **Choose the Google Voice Typing item.**

4. **Disable the Block Offensive Words option.**

And just what are offensive words? I would think that *censorship* is an offensive word. But no. Apparently, only a few choice words fall into this category. I won't print them here, because the phone's censor retains the initial letter and generally makes the foul language easy to guess. D***.

Add a Word to the Dictionary

Betcha didn't know that your phone sports a dictionary. The dictionary keeps track of words you type — words that it may not recognize as being spelled properly.

Words unknown to the phone are highlighted on the screen. Sometimes, the word is shown in a different color or on a different background or even underlined in red. To add that word to the phone's dictionary, long-press it. You see the Add Word to Dictionary command, which sticks the word in the phone's dictionary.

To review or edit the phone's dictionary, follow these steps:

1. **Start the Settings app.**

2. **Choose Language & Input.**

3. **Choose Personal Dictionary.**

   The command may not be obvious on some phones: Try choosing the keyboard first, and then choose either the Dictionary or User Dictionary command.

When the dictionary is visible, you can review words, edit them, remove them, or manually add new ones. To edit or delete a word, long-press it. To add a word, manually tap the Add icon.

Employ Some Useful Widgets

Your phone features a wide assortment of widgets with which to festoon your phone's Home screen. They can be exceedingly handy, although you may not realize it because the sample widgets often included with the phone are tepid and lack inspiration.

Good widgets to add include navigation, contact, eBook, and web page favorites. Adding any of these widgets starts out the same. Here are the brief directions, with more specifics offered in Chapter 19:

1. **Long-press a Home screen page that has room for the widget.**
2. **Choose Widgets.**
3. **Drag the desired widget to the Home screen.**
4. **Complete the process.**

   The process is specific for each type of widget suggestion in this section.

Direct Dial widget

Those phone numbers you dial all the time are accessed most quickly by choosing the Direct Dial widget. After adding the widget, choose a contact from the phone's address book. For contacts with multiple numbers, choose a number (all are displayed in the list). Tapping this widget instantly dials the number.

Directions widget

The Directions widget allows you to rapidly summon directions to a specific location from wherever you happen to be. After you plop the widget on the Home screen, select a traveling method and destination. You can type a contact name, an address, a business name, and so on. Add a shortcut name, which is a brief description to fit under the widget on the Home screen. Tap the Save button.

Tap the Directions widget to use it. Instantly, the Maps app starts and enters Navigation mode, steering you from wherever you are to the location referenced by the widget.

eBook widget

When you're mired in the middle of that latest potboiler, put a Book shortcut on the Home screen: Choose the Book widget, and then browse your digital library for the eBook you want to access. Tap the widget to open the Play Books app and jump right to the spot where you were last reading.

Web page bookmark widget

While slapping down the Bookmark widget would be an obvious way to keep handy a favorite web page shortcut on the Home screen, I have an easier way: Open the web browser app, navigate to the page you desire, and then tap the Action Overflow and choose Add to Homescreen.

Take a Screen Shot

A *screen shot,* also called a *screen cap* (for *cap*ture), is a picture of your phone's touchscreen. So if you see something interesting on the screen, or if you just want to take a quick pic of your phone life, take a screen shot.

The stock Android method of shooting the screen is to press and hold both the Volume Down and Power/Lock keys at the same time. Upon success, the touchscreen image reduces in size, you may hear a shutter sound, and the screen shot is saved.

- Open the Photos app to view your screen shots.

- Some Samsung phones use a motion action to capture the screen: Hold your hand perpendicular to the phone, like you're giving it a karate chop. Swipe the edge of your palm over the screen, from right to left or from left to right. Upon success, you hear a shutter sound. The screen is captured.

- Internally, screen shots are stored in the Pictures/Screenshots folder. They're created in the PNG graphics file format.

25

Ten Things to Remember

*I*f only it were easy to narrow to ten items the list of all the things I want you to remember when using your Android phone. Even though you'll find in this chapter ten good things not to forget, don't think for a moment that there are *only* ten. In fact, as I remember more, I'll put them on my website, at www.wambooli.com. Check it for updates about your phone and perhaps for even more things to remember.

Lock the Phone on a Call

Whether you dialed out or someone dialed in, after you start talking, lock your phone. Press the Power/Lock key. By doing so, you disable the touchscreen and ensure that the call isn't unintentionally disconnected.

Of course, the call can still be disconnected by a dropped signal or by the other party getting all huffy and hanging up on you. But by locking the phone, you prevent a stray finger or your pocket from disconnecting (or muting) the phone.

If you like to talk with your hands, or just use your hands while you're on the phone (I sweep the floor, for example), get a good set of earbuds with a microphone. Using a headset lets you avoid trying to hold the phone between your ear and shoulder, which could unlock the phone or cause you to drop it or perhaps do something more perilous.

Switch to Landscape Orientation

The natural orientation of the Android phone is vertical — its *portrait* orientation. Even so, that doesn't mean you have to use an app in portrait orientation.

Rotating the phone to its side makes many apps, such as the Web Browser app and the Play Books app, appear wider. It's often a better way to see things in the phone's landscape orientation. For typing, that orientation gives you larger key caps on which to type when you use the onscreen keyboard.

✔ Not every app supports landscape orientation.

✔ The phone's Home screen may not support landscape orientation.

✔ You can lock the orientation so that the touchscreen doesn't flip and flop. The easiest way to do so is by tapping the Auto Rotate quick setting. See Chapter 3 for information on the Quick Settings.

Dictate Text

Dictation is such a handy feature — don't forget to use it! You can dictate most text instead of typing it. Especially for text messages, it's just so quick and handy: Tap the Dictation (Microphone) key on the onscreen keyboard. Actually, anywhere you see the Dictation icon (shown in the margin), tap and begin speaking. Your utterances are translated to text. In most cases, the translation is instantaneous.

✔ See Chapter 4 for more information on dictation.

✔ Google Now doesn't require you to tap the Microphone icon. Instead, utter the phrase "OK Google" and it starts listening. See Chapter 15 for information about Google Now.

Enjoy Predictive Text

Don't forget to take advantage of the suggestions that appear above the onscreen keyboard while you type. Tap the Space key or a word suggestion to select and "type" that word. Plus, the predictive-text feature may instantly display the next logical word for you.

When predictive text fails you, keep in mind that you can use Google Gesture typing instead of the old hunt-and-peck. Dragging your finger over the keyboard and then choosing a word suggestion works quickly — when you remember to do it.

See Chapter 20 for information on activating these keyboard settings.

Avoid Battery Hogs

Three items on your phone suck down battery power faster than an 18-year-old fleeing the tyranny of high school on graduation day:

- The display
- Navigation
- Wireless radios

The display is obviously a most necessary part of your phone, but it's also a tremendous power hog. The Adaptive Brightness (also called Auto Brightness) setting is your best friend for saving power with the display. See Chapter 20.

Navigation is certainly handy, but the battery drains rapidly because the phone's touchscreen is on the entire time, the voice is dictating directions, and the Wi-Fi and GPS radios may be active. If possible, try to plug the phone into the car's power socket while you're navigating. If you can't, keep a sharp eye on the battery meter.

Bluetooth requires extra power for its wireless radio. When you need that level of connectivity, great! Otherwise, turn off your Bluetooth gizmo as soon as necessary to save power.

Wireless radios include Wi-Fi networking, Bluetooth, and GPS. Though they do require extra power, they aren't power hogs, like navigation and the display. Still, when power is getting low, consider disabling those items.

See Chapter 23 for more information on managing the phone's battery.

Beware of Roaming

Roaming can be expensive. The last non-smartphone (dumbphone?) I owned racked up $180 in roaming charges the month before I switched to a better cellular plan. Even though you might have a good phone plan, keep an eye on the phone's status bar to ensure that you don't see the Roaming status icon when you're making a call.

Well, yes, it's okay to make a call when your phone is roaming. My advice is to remember to *check* for the icon, not to avoid it. If possible, try to make your phone calls when you're back in your cellular service's coverage area. If you can't, make the phone call but keep in mind that you will be charged roaming fees. They ain't cheap.

Properly Use the Plus (+) Symbol When Dialing

That phone number may look like it needs the + symbol, and the Phone app's dialpad features a + key, shared with the 0 key, but don't do it! The + symbol is used for dialing international numbers. It is not a separator symbol.

The proper procedure is to start an international phone number with the + character, followed by the country code. Then comes the rest of the number. See Chapter 22 for more information on international dialing.

Check Your Schedule

The Calendar app reminds you of upcoming dates and generally keeps you on schedule. A great way to augment the calendar is to employ the Calendar widget on the Home screen.

The Calendar widget lists the current date and then a long list of upcoming appointments. It's a great way to check your schedule, especially when you're a busy person and you occasionally forget to show up where you're needed.

See Chapter 20 for information on adding widgets to the Home screen; Chapter 15 covers the Calendar app.

Snap a Pic of That Contact

Here's something I always forget: Whenever you're near one of your contacts, take the person's picture. Sure, some people are bashful, but most folks are flattered. The idea is to build up the phone's Address Book app so that all contacts have photos. Receiving a call is then much more interesting when you see the caller's picture, especially a silly or an embarrassing one.

When taking the picture, be sure to show it to the person before you assign it to the contact. Let them decide whether it's good enough. Or if you just want to be rude, assign a crummy-looking picture. Heck, you don't even have to do that: Just assign a random picture of anything. A plant. A rock. Your dog. But seriously, the next time you meet up with a contact, keep in mind that the phone can take that person's picture.

See Chapter 13 for more information on using the phone's camera.

Use the Search Icon

Google is synonymous with searching, so you would expect the Google phone to sport robust searching abilities. And it does!

The Search icon is found in just about every Android app and on almost every screen on your phone. Tap that icon to look for information such as locations, people, text — you name it. It's handy. It's everywhere. Use it.

Once upon a time, the Search icon was one of the standard navigation icons that appeared at the bottom of an Android phone's touchscreen. The original four navigation icons were Home, Back, Menu (now Action Overflow), and Search.

Index

● *T* ●

• •

• Z •

About the Author

Dan Gookin has been writing about technology for over 25 years. He combines his love of writing with his gizmo fascination to create books that are informative, entertaining, and not boring. Having written over 130 titles with 12 million copies in print translated into over 30 languages, Dan can attest that his method of crafting computer tomes seems to work.

Perhaps his most famous title is the original *DOS For Dummies,* published in 1991. It became the world's fastest-selling computer book, at one time moving more copies per week than the *New York Times* number-one bestseller (though, as a reference, it could not be listed on the *Times'* Best Sellers list). That book spawned the entire line of *For Dummies* books, which remains a publishing phenomenon to this day.

Dan's most popular titles include *PCs For Dummies, Word For Dummies, Laptops For Dummies,* and *Android Phones For Dummies*. He also maintains the vast and helpful website www.wambooli.com.

Dan holds a degree in Communications/Visual Arts from the University of California, San Diego. He lives in the Pacific Northwest, where he enjoys spending time with his sons playing video games indoors while they enjoy the gentle woods of Idaho.

Publisher's Acknowledgments

Acquisitions Editor: Katie Mohr

Senior Project Editor: Paul Levesque

Copy Editor: Rebecca Whitney

Editorial Assistant: Claire Brock

Sr. Editorial Assistant: Cherie Case

Project Coordinator: Antony Sami

Cover Image: tanuha2001/Shutterstock, Sam72/Shutterstock